SpringerBriefs in Physics

SpringerBriefs in Physics are a series of slim high-quality publications encompassing the entire spectrum of physics. Manuscripts for SpringerBriefs in Physics will be evaluated by Springer and by members of the Editorial Board. Proposals and other communication should be sent to your Publishing Editors at Springer.

Featuring compact volumes of 50 to 125 pages (approximately 20,000–45,000 words), Briefs are shorter than a conventional book but longer than a journal article. Thus, Briefs serve as timely, concise tools for students, researchers, and professionals. Typical texts for publication might include:

- A snapshot review of the current state of a hot or emerging field
- A concise introduction to core concepts that students must understand in order to make independent contributions
- An extended research report giving more details and discussion than is possible in a conventional journal article
- A manual describing underlying principles and best practices for an experimental technique
- An essay exploring new ideas within physics, related philosophical issues, or broader topics such as science and society

Briefs allow authors to present their ideas and readers to absorb them with minimal time investment. Briefs will be published as part of Springer's eBook collection, with millions of users worldwide. In addition, they will be available, just like other books, for individual print and electronic purchase. Briefs are characterized by fast, global electronic dissemination, straightforward publishing agreements, easy-to-use manuscript preparation and formatting guidelines, and expedited production schedules. We aim for publication 8–12 weeks after acceptance.

More information about this series at http://www.springer.com/series/8902

Georgios Pastras

The Weierstrass Elliptic Function and Applications in Classical and Quantum Mechanics

A Primer for Advanced Undergraduates

Springer

Georgios Pastras
Institute of Nuclear and Particle Physics
National Centre for Scientific Research
"Demokritos"
Agia Paraskevi, Greece

ISSN 2191-5423   ISSN 2191-5431   (electronic)
SpringerBriefs in Physics
ISBN 978-3-030-59384-1   ISBN 978-3-030-59385-8   (eBook)
https://doi.org/10.1007/978-3-030-59385-8

This Springer imprint is published by the registered company Springer Nature Switzerland AG
The registered company address is: Gewerbestrasse 11, 6330 Cham, Switzerland

*Dedicated to the memory of my father
Konstantinos*

Preface

The field of elliptic functions, apart from its own mathematical beauty, has many applications in physics in a variety of topics, such as string theory or integrable systems. Most physics or applied mathematics students, who desire to explore the field, do not have available time for a full mathematics course. As a consequence, they have to consult either handbooks of formulae, which do not provide deep knowledge on the subject, or textbooks for abstract mathematicians, which are written in a non-familiar language.

This text aims at senior undergraduates in physics or applied mathematics. Of course, junior graduate students, who are not familiar with the subject of elliptic functions, will also benefit from this book. The text focuses on the Weierstrass theory of elliptic functions instead of the Jacobi theory. This is a personal preference of the author. At a deeper level this preference originates from the fact the the Weierstrass elliptic function uniformizes the torus, when expressed as an elliptic curve.

The text provides a fast, but thorough introduction to the mathematical theory and then presents some important applications in classical and quantum physics. I expect the reader to benefit from the structure of this book, based on the following principle: For a physicist or an applied mathematician, mathematics and physics should form an endless loop; Better knowledge of mathematics means more tools to do physics. At the same time, the intuition gained by the solution of physical problems leads to better understanding of the mathematical tools. Following this, I believe the simple applications, such as the particle in a cubic potential or the simple pendulum, will greatly help the reader to develop physical intuition on the behaviour of the Weierstrass elliptic and related functions.

At the last chapters of the book, some more interesting examples are presented, such as the $n = 1$ Lamé problem. Everyone is aware of Bloch's theorem, however, very few undergraduates have ever seen a periodic potential where they can specify the band structure analytically. This can be performed in the case of the $n = 1$ Lamé problem, with the help of elliptic functions. The text is supplemented by problems and solutions to help the reader gain familiarity not only with the notions, but also with the techniques. The advanced reader will find some more difficult and interesting problems, which are inspired by contemporary research on classical string solutions and the sine-Gordon equation, at the last chapter.

For Chaps. 1–3, the reader is required to have basic knowledge on complex calculus including Cauchy's residue theorem. For Chap. 4, elementary experience on classical mechanics is required, mainly on one-dimensional problems and the conservation of energy. Finally, for Chap. 5, the reader is required to be familiar with basic quantum mechanics, including Schrödinger's equation, while some familiarity with Bloch's theorem will also be helpful.

I wish to everyone, who desires to explore and conquer the seductive relation between mathematics and physics, to find this book enjoyable.

Athens, Greece   Georgios Pastras
July 2020

Acknowledgements

Although he is no longer with us, I would like to thank my mentor Ioannis Bakas, who introduced me to the subject of the Weierstrass elliptic function long ago. It was a short but very well-aimed introduction, consistent with his temperament in physics.

I would like to thank Eleftherios Papantonopoulos, who organized the series of lectures that I presented at the School of Applied Mathematics and Physical Sciences of the National Technical University of Athens in May 2017. The notes for these lectures became the core of this book.

Finally, I would like to thank my close collaborators Dimitrios Katsinis and Ioannis Mitsoulas for proofreading and their useful comments.

Contents

Chapter 1
The Weierstrass Elliptic Function

Abstract The elliptic functions are meromorphic complex functions, which are periodic in two distinct directions in the complex plane. As such, they are naturally well-defined on the torus. For this reason, they find numerous applications in physics. Although their definition is quite simple, it leads to a particularly rich and beautiful set of properties. In this chapter, we study the generic properties of elliptic functions. Then, we proceed to construct the Weierstrass elliptic function and study its features, focusing on those that are particularly useful for its applications in Physics.

1.1 Elliptic Functions

The elliptic functions are defined in a very simple way: they are doubly-periodic meromorphic functions of one complex variable. Although their definition is quite simple, it leads to a particularly rich and beautiful set of properties. It is not an exaggeration to say that a detailed and thorough study of these properties allows an understanding of the behaviour of the elliptic functions, as intuitive as the one that everyone has on the behaviour of the simple trigonometric functions.

Historically, the elliptic functions were introduced by Abel in his try to invert the elliptic integrals. The first original constructions of elliptic functions are due to Weierstrass [1] and Jacobi [2]. In this book, we focus on the former. The two constructions are equivalent; it is quite simple to express the Weierstrass's elliptic function in terms of Jacobi's elliptic functions and vice versa.

Excellent pedagogical texts on the subject of elliptic functions are the classic text by Watson and Whittaker [3] and the more specialized text by Akhiezer [4]. Useful reference handbooks with many details on transcendental functions, including those used in this book, are provided by Bateman and Erdélyi, [5], which is freely available online, as well as the classical reference by Abramowitz and Stegun [6].

© The Author(s), under exclusive license to Springer Nature Switzerland AG 2020
G. Pastras, *The Weierstrass Elliptic Function and Applications in Classical and Quantum Mechanics*, SpringerBriefs in Physics,
https://doi.org/10.1007/978-3-030-59385-8_1

1.1.1 Basic Definitions

Let us consider a complex function of one complex variable $f(z)$ obeying the property

$$f(z + 2\omega_1) = f(z), \quad f(z + 2\omega_2) = f(z), \tag{1.1}$$

for two complex numbers ω_1 and ω_2, whose ratio is not purely real (thus, they correspond to different directions on the complex plane). Such a function is called *doubly-periodic* with periods $2\omega_1$ and $2\omega_2$. A meromorphic, doubly-periodic function is called an *elliptic* function.

The complex numbers 0, $2\omega_1$, $2\omega_2$ and $2\omega_1 + 2\omega_2$ define a parallelogram on the complex plane. Knowing the values of the elliptic function within this parallelogram completely determines the elliptic function, as a consequence of the property (1.1). However, instead of $2\omega_1$ and $2\omega_2$, one could use any pair of linear combinations of the latter with integer coefficients, provided that their ratio is not real. In the general case, the aforementioned parallelogram can be divided to several identical cells. If $2\omega_1$ and $2\omega_2$ have been selected to be "minimal", i.e. if there is no 2ω within the parallelogram (boundaries included, vertices excepted), such that $f(z + 2\omega) = f(z)$, then the periods $2\omega_1$ and $2\omega_2$ are called *fundamental periods* and the parallelogram is called a *fundamental period parallelogram*.

Two points z_1 and z_2 on the complex plane whose difference is an integer multiple of the periods

$$z_2 - z_1 = 2m\omega_1 + 2n\omega_2, \quad m, n \in \mathbb{Z}, \tag{1.2}$$

are called *congruent* to each other. For such points we will use the notation

$$z_1 \sim z_2 \Leftrightarrow z_2 - z_1 = 2m\omega_1 + 2n\omega_2, \quad m, n \in \mathbb{Z}. \tag{1.3}$$

Obviously, by definition, the elliptic function at congruent points assumes the same value,

$$z_1 \sim z_2 \Rightarrow f(z_1) = f(z_2). \tag{1.4}$$

A parallelogram defined by the points z_0, $z_0 + 2\omega_1$, $z_0 + 2\omega_2$ and $z_0 + 2\omega_1 + 2\omega_2$, for any z_0, is called a "*cell*" (Fig. 1.1). It is often useful to use the boundary of an arbitrary cell instead of the fundamental period parallelogram to perform contour integrals, when poles appear at the boundary of the latter.

Knowing the roots and poles of an elliptic function within a cell suffices to describe all roots and poles of the elliptic function, as all other roots and poles are congruent to the former. As such, a set of roots and poles congruent to those within a cell is called an *irreducible set of roots* or an *irreducible set of poles*, respectively.

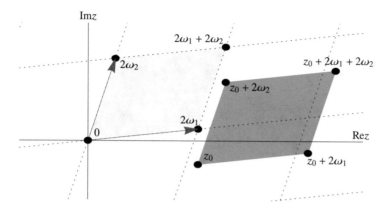

Fig. 1.1 The fundamental period parallelogram is shaded with blue and an arbitrary cell is shaded with red. The dashed lines define the period parallelograms

1.1.2 Modular Transformations

Given two fundamental periods $2\omega_1$ and $2\omega_2$, one can define two different periods as,

$$\omega_1' = a\omega_1 + b\omega_2, \tag{1.5}$$

$$\omega_2' = c\omega_1 + d\omega_2, \tag{1.6}$$

where $a, b, c, d \in \mathbb{Z}$. Any period in the lattice defined by ω_1' and ω_2', namely $2\omega = 2m'\omega_1' + 2n'\omega_2'$ is obviously a period of the old lattice, but is the opposite also true? i.e. are the new periods ω_1' and ω_2' a fundamental pair of periods? In order for the opposite statement to hold, the area of the period parallelogram defined by the new periods $2\omega_1'$ and $2\omega_2'$ has to be equal to the area of the fundamental period parallelogram defined by the original ones $2\omega_1$ and $2\omega_2$, (see Fig. 1.2). The area of the parallelogram defined by two complex numbers z_1 and z_2 is given by

$$A = |\mathrm{Im}\,(z_1 \bar{z}_2)|. \tag{1.7}$$

It is a matter of simple algebra to show that

$$\mathrm{Im}\left(\omega_1' \bar{\omega}_2'\right) = (ad - bc)\,\mathrm{Im}\left(\omega_1 \bar{\omega}_2\right).$$

Thus, the new periods can generate the while original lattice if

$$\begin{vmatrix} a & b \\ c & d \end{vmatrix} = \pm 1, \tag{1.8}$$

in other words, if

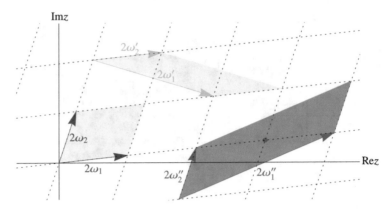

Fig. 1.2 The green periods $2\omega_1' = 4\omega_1 - 2\omega_2$ and $2\omega_2' = 2\omega_1$ can generate the whole original lattice. The red periods $2\omega_1'' = 4\omega_1 + 2\omega_2$ and $2\omega_2'' = 2\omega_2$ generate only half of the original lattice. For example, there is no way to reach the red-dotted vertex by a linear combination of $2\omega_1''$ and $2\omega_2''$ with integer coefficients. The area of the parallelogram defined by $2\omega_1'$ and $2\omega_2'$ is equal to the area of the parallelogram defined by the original periods, whereas the one defined by $2\omega_1''$ and $2\omega_2''$ has double this area

$$\begin{pmatrix} a & b \\ c & d \end{pmatrix} \in SL\,(2,\mathbb{Z})\,. \tag{1.9}$$

It is a direct consequence that an elliptic function necessarily obeys

$$f\,(z;\omega_1,\omega_2) = f\left(z;\omega_1',\omega_2'\right), \tag{1.10}$$

when

$$\begin{pmatrix} \omega_1' \\ \omega_2' \end{pmatrix} = S \begin{pmatrix} \omega_1 \\ \omega_2 \end{pmatrix}, \tag{1.11}$$

where $S \in SL\,(2,\mathbb{Z})$.

1.1.3   Basic Properties of Elliptic Functions

Before we proceed to the specific construction of an elliptic function, it is wise to study some of the generic properties of elliptic functions. This is going to be useful in multiple ways. Firstly, these properties are going to hint what are the simplest possible specific realizations of an elliptic function. Secondly, when we are going to have a specific elliptic function in hand, it will be possible to distinguish which of its properties are direct consequences of the fact that it is an elliptic function and which are unique for the specific realization.

Theorem 1.1 *The sum of residues over an irreducible set of poles of an elliptic function vanishes.*

To demonstrate this, we use Cauchy's residue theorem over the boundary of a cell. By definition, the poles of an elliptic function within a cell is an irreducible set of poles. Therefore,

$$\sum_{z_i \in \text{irreducible set of poles}} \text{Res}\,(f, z_i) = \frac{1}{2\pi i} \oint_{\partial \text{cell}} f\,(z)\,dz$$

$$= \frac{1}{2\pi i} \int_{z_0}^{z_0+2\omega_1} f\,(z)\,dz + \frac{1}{2\pi i} \int_{z_0+2\omega_1}^{z_0+2\omega_1+2\omega_2} f\,(z)\,dz$$

$$+ \frac{1}{2\pi i} \int_{z_0+2\omega_1+2\omega_2}^{z_0+2\omega_2} f\,(z)\,dz + \frac{1}{2\pi i} \int_{z_0+2\omega_2}^{z_0} f\,(z)\,dz.$$

Shifting z by $2\omega_1$ in the second integral and by $2\omega_2$ in the third, yields

$$\sum_{z_i \in \text{irreducible set of poles}} \text{Res}\,(f, z_i) = \frac{1}{2\pi i} \int_{z_0}^{z_0+2\omega_1} [f\,(z) - f\,(z + 2\omega_2)]\,dz$$

$$- \frac{1}{2\pi i} \int_{z_0}^{z_0+2\omega_2} [f\,(z) - f\,(z + 2\omega_1)]\,dz,$$

which vanishes as a consequence of f being a doubly periodic function with periods $2\omega_1$ and $2\omega_2$. Therefore,

$$\sum_{z_i \in \text{irreducible set of poles}} \text{Res}\,(f, z_i) = 0. \tag{1.12}$$

Theorem 1.2 *An elliptic function with an empty irreducible set of poles is a constant function.*

An elliptic function with no poles in a cell, necessarily has no poles at all, as a pole outside a cell necessarily would have a congruent pole within the cell. Consequently, such a function is not just meromorphic, but rather it is analytic. Furthermore, an analytic function in a cell is necessarily bounded within the cell. A direct consequence of property (1.1) is that an analytic elliptic function is bounded everywhere. But a bounded analytic function is necessarily a constant function.

The number of roots of the equation

$$f\,(z) = z_0 \tag{1.13}$$

within a cell is the same for all $z_0 \in \mathbb{C}$. Before demonstrating this, we will review some properties of Cauchy's integral.

Consider a meromorphic function g, with a number of poles u_i and roots w_j with multiplicities r_i and s_j respectively, within a region bounded by a closed contour C.

The Laurent series of the function g at the regime of a pole or a root z_0 is

$$g(z) \simeq c_m(z - z_0)^m + c_{m+1}(z - z_0)^{m+1} + \cdots$$

with $c_m \neq 0$. In the case z_0 is a root w_j, then $m = s_j > 0$, while in the case z_0 is a pole u_i, then $m = -r_i < 0$. The derivative of g in the regime of a pole or root is

$$g'(z) \simeq m c_m(z - z_0)^{m-1} + (m + 1) c_{m+1}(z - z_0)^m + \cdots$$

Furthermore, consider an analytic function h. Its Laurent series at the regime of a pole or root of g is trivially

$$h(z) \simeq h(z_0) + h'(z_0)(z - z_0) + \cdots$$

It is a matter of simple algebra to show that the Laurent series of the function hg'/g at the region of z_0 is

$$h(z) \frac{g'(z)}{g(z)} \simeq \frac{mh(z_0)}{z - z_0} + mh'(z_0) + c_{m+1}\left(m + 1 - \frac{1}{c_m^2}\right)h(z_0) + \mathcal{O}(z - z_0)^2.$$

Thus, at any root or pole of the function g, the function hg'/g has a first order pole with residue $mh(z_0)$. It is a direct consequence of Cauchy's residue theorem that

$$\frac{1}{2\pi i} \oint_C h(z) \frac{g'(z)}{g(z)} = \sum_j s_j h(w_j) - \sum_i r_i h(u_i). \tag{1.14}$$

Let's now return to the case of an elliptic function f. We would like to calculate the contour integral of formula (1.14) with $h(z) = 1$, $g(z) = f(z) - z_0$ and C being the boundary of a cell, namely

$$I = \frac{1}{2\pi i} \oint_{\partial \text{cell}} \frac{f'(z)}{f(z) - z_0}.$$

The function $f - z_0$ is trivially elliptic, while differentiating equations (1.1), one yields

$$f'(z + 2\omega_1) = f'(z), \quad f'(z + 2\omega_2) = f'(z), \tag{1.15}$$

implying that f' is also an elliptic function with the same periods as f. In an obvious manner, the function $f'(z)/(f(z) - z_0)$ is an elliptic function with the same periods as f. A direct application of the fact that the sum of the residues of an elliptic function over a cell vanishes (1.12) is

$$I = \frac{1}{2\pi i} \oint_{\partial \text{cell}} \frac{f'(z)}{f(z) - z_0} = 0.$$

This combined with Eq. (1.14) yields

$$\sum_j s_j - \sum_i r_i = 0. \qquad (1.16)$$

Therefore,

Theorem 1.3 *The number of roots of the equation $f(z) = z_0$ in a cell is equal to the number of poles of f in a cell (both weighted by their multiplicity), independently of the value of z_0.*

This number is called the *order* of the elliptic function f.

Theorem 1.4 *The order of a non-constant elliptic function cannot be equal to 1.*

A non-constant elliptic function has necessarily at least one pole in a cell as a consequence of Theorem 1.2, thus its order is at least 1. However, an elliptic function of order 1 necessarily has only a single first order pole in a cell. In such a case though, the sum of the residues of the elliptic function in a cell equals to the residue of this single pole, and, thus, it cannot vanish. This contradicts (1.12) and therefore an elliptic function cannot be of order one. The lowest order elliptic functions are of order 2. Such a function can have either a single second order pole or two first order poles with opposite residues, in any cell.

Theorem 1.5 *The sum of the locations of an irreducible set of poles (weighted by their multiplicity) is congruent to the sum of the locations of an irreducible set of roots (also weighted by their multiplicity).*

To demonstrate this, we will calculate Cauchy's integral with $h(z) = z$, $g(z) = f(z)$ and as contour of integration C the boundary of a cell. The left hand side of (1.14) equals

$$I = \frac{1}{2\pi i} \oint_{\partial \mathrm{cell}} \frac{z f'(z)}{f(z)}$$

$$= \frac{1}{2\pi i} \int_{z_0}^{z_0+2\omega_1} \frac{z f'(z)}{f(z)} dz + \frac{1}{2\pi i} \int_{z_0+2\omega_1}^{z_0+2\omega_1+2\omega_2} \frac{z f'(z)}{f(z)} dz$$

$$+ \frac{1}{2\pi i} \int_{z_0+2\omega_1+2\omega_2}^{z_0+2\omega_2} \frac{z f'(z)}{f(z)} dz + \frac{1}{2\pi i} \int_{z_0+2\omega_2}^{z_0} \frac{z f'(z)}{f(z)} dz.$$

We shift z by $2\omega_1$ in the second integral and by $2\omega_2$ in the third one and we get

$$I = \frac{1}{2\pi i} \int_{z_0}^{z_0+2\omega_1} \left(\frac{z f'(z)}{f(z)} - \frac{(z+2\omega_2) f'(z+2\omega_2)}{f(z+2\omega_2)} \right) dz$$

$$- \frac{1}{2\pi i} \int_{z_0}^{z_0+2\omega_2} \left(\frac{z f'(z)}{f(z)} - \frac{(z+2\omega_1) f'(z+2\omega_1)}{f(z+2\omega_1)} \right) dz.$$

Using the periodicity properties of f and f', (1.1) and (1.15) respectively, yields

$$
\begin{aligned}
I &= \frac{1}{2\pi i} \int_{z_0}^{z_0+2\omega_1} \left(\frac{zf'(z)}{f(z)} - \frac{(z+2\omega_2) f'(z)}{f(z)} \right) dz \\
&\quad - \frac{1}{2\pi i} \int_{z_0}^{z_0+2\omega_2} \left(\frac{zf'(z)}{f(z)} - \frac{(z+2\omega_1) f'(z)}{f(z)} \right) dz \\
&= -\frac{\omega_2}{\pi i} \int_{z_0}^{z_0+2\omega_1} \frac{f'(z)}{f(z)} dz + \frac{\omega_1}{\pi i} \int_{z_0}^{z_0+2\omega_2} \frac{f'(z)}{f(z)} dz \\
&= -\frac{\omega_2}{\pi i} \left(\ln f(z_0+2\omega_1) - \ln f(z_0) \right) + \frac{\omega_1}{\pi i} \left(\ln f(z_0+2\omega_2) - \ln f(z_0) \right).
\end{aligned}
$$

Although $z_0 + 2\omega_1 \sim z_0 \sim z_0 + 2\omega_2$, due to the branch cut of the logarithmic function, in general we have that $\ln f(z_0+2\omega_1) - \ln f(z_0) = 2im\pi$ and $\ln f(z_0+2\omega_2) - \ln f(z_0) = 2in\pi$, with $m, n \in \mathbb{Z}$. Thus,

$$
I = 2m\omega_1 + 2n\omega_2 \sim 0.
$$

Finally, applying property (1.14) we get

$$
I = \sum_i r_i u_i - \sum_j s_j w_j,
$$

which implies that

$$
\sum_j r_j u_j \sim \sum_i s_i w_i, \tag{1.17}
$$

which concludes the proof of Theorem 1.5.

1.2 The Weierstrass Elliptic Function

As we showed in previous section, the lowest possible order of an elliptic function is 2. This leads to two possible paths to follow. The first path is the construction of an elliptic function with two first order poles with opposite residues at each cell. This path leads to Jacobi's elliptic functions. The other path is the construction of an elliptic function with a single second order pole in each cell. This path leads to Weierstrass's elliptic function. In this book we follow the latter path.

1.2.1 Definition

The construction of an elliptic function with a single second order pole in each cell is not a difficult task. It suffices to sum an infinite set of copies of the function $1/z^2$

each one shifted by $2m\omega_1 + 2n\omega_2$ for all $m, n \in \mathbb{Z}$. The usual convention includes the addition of a constant cancelling the contributions of all these functions at $z = 0$ (except for the term with $m = n = 0$), so that the Laurent series of the constructed function at the region of $z = 0$ has a vanishing zeroth order term. Following these directions, we define,

$$\wp(z) := \frac{1}{z^2} + \sum_{\{m,n\}\neq\{0,0\}} \left(\frac{1}{(z + 2m\omega_1 + 2n\omega_2)^2} - \frac{1}{(2m\omega_1 + 2n\omega_2)^2} \right). \quad (1.18)$$

By construction, this function is doubly periodic with fundamental periods equal to $2\omega_1$ and $2\omega_2$.

$$\wp(z + 2\omega_1) = \wp(z), \quad \wp(z + 2\omega_2) = \wp(z). \quad (1.19)$$

This function is called *Weierstrass elliptic function*.

1.2.2 Basic Properties

A direct consequence of the definition (1.18) is the fact that the Weierstrass elliptic function is an even function

$$\wp(-z) = \wp(z). \quad (1.20)$$

Let's acquire the Laurent series of the Weierstrass elliptic function at the regime of $z = 0$. It is easy to show that

$$\frac{1}{(z + w)^2} - \frac{1}{w^2} = \frac{1}{w^2} \sum_{k=1}^{\infty} (k + 1) \left(-\frac{z}{w} \right)^k.$$

Consequently,

$$\wp(z) = \frac{1}{z^2} + \sum_{k=1}^{\infty} \sum_{\{m,n\}\neq\{0,0\}} \frac{(k + 1)(-1)^k}{(m\omega_1 + n\omega_2)^{k+2}} z^k = \frac{1}{z^2} + \sum_{k=1}^{\infty} a_k z^k,$$

where

$$a_k = (k + 1)(-1)^k \sum_{\{m,n\}\neq\{0,0\}} \frac{1}{(2m\omega_1 + 2n\omega_2)^{k+2}}.$$

The fact that $\wp$ is even implies that only the even indexed coefficients do not vanish,

$$a_{2l+1} = 0, \quad (1.21)$$

$$a_{2l} = (2l + 1) \sum_{\{m,n\}\neq\{0,0\}} \frac{1}{(2m\omega_1 + 2n\omega_2)^{2(l+1)}}. \quad (1.22)$$

For reasons that will become apparent later, we define g_2 and g_3 so that

$$\wp(z) = \frac{1}{z^2} + \frac{g_2}{20}z^2 + \frac{g_3}{28}z^4 + \mathcal{O}\left(z^6\right), \tag{1.23}$$

implying that

$$g_2 := 60 \sum_{\{m,n\}\neq\{0,0\}} \frac{1}{(2m\omega_1 + 2n\omega_2)^4}, \tag{1.24}$$

$$g_3 := 140 \sum_{\{m,n\}\neq\{0,0\}} \frac{1}{(2m\omega_1 + 2n\omega_2)^6}. \tag{1.25}$$

1.2.3   Weierstrass Differential Equation

By direct differentiation of Eq. (1.18), we express the derivative of Weierstrass function as

$$\wp'(z) = -2 \sum_{m,n} \frac{1}{(z + m\omega_1 + n\omega_2)^3}. \tag{1.26}$$

It follows that the derivative of the Weierstrass elliptic function is an odd function

$$\wp'(-z) = -\wp'(z). \tag{1.27}$$

The Laurent series of $\wp(z)$, $\wp'(z)$, $\wp^3(z)$ and $\wp'^2(z)$ at the regime of $z = 0$ are

$$\wp(z) = \frac{1}{z^2} + \frac{g_2}{20}z^2 + \frac{g_3}{28}z^4 + \mathcal{O}\left(z^6\right),$$

$$\wp'(z) = -\frac{2}{z^3} + \frac{g_2}{10}z + \frac{g_3}{7}z^3 + \mathcal{O}\left(z^5\right),$$

$$\wp^3(z) = \frac{1}{z^6} + \frac{3g_2}{20}\frac{1}{z^2} + \frac{3g_3}{28} + \mathcal{O}\left(z^2\right),$$

$$\wp'^2(z) = \frac{4}{z^6} - \frac{2g_2}{5}\frac{1}{z^2} - \frac{4g_3}{7} + \mathcal{O}\left(z^2\right).$$

Observing these expressions, we realize that there is a linear combination of the above, which is not singular at $z = 0$. One can eliminate the sixth order pole by taking an appropriate combination of $\wp'^2$ and $\wp^3$. This leaves a function with a second order pole. Taking an appropriate combination of the latter and $\wp$ allows to write down a function with no poles at $z = 0$. Trivially, adding an appropriate constant results in a non-singular function, which is also vanishing at $z = 0$. The appropriate combination turns out to be

$$\wp'^2 (z) - 4\wp^3 (z) + g_2\wp (z) + g_3 = \mathcal{O}\left(z^2\right).$$

But, the derivative, as well as powers of an elliptic function are elliptic functions with the same periods. Therefore, the function $\wp'^2 (z) - 4\wp^3 (z) + g_2\wp (z) + g_3$ is an elliptic function with the same periods as $\wp (z)$. Since the latter has no pole at $z = 0$, it does not have any pole in the fundamental period parallelogram, and, thus, it does not have any pole at all; it is an elliptic function with no poles. According to Theorem 1.2, elliptic functions with no poles are necessarily constants and since $\wp'^2 (z) - 4\wp^3 (z) + g_2\wp (z) + g_3$ vanishes at the origin, it vanishes everywhere. This implies that the Weierstrass elliptic function obeys the differential equation,

$$\wp'^2 (z) = 4\wp^3 (z) - g_2\wp (z) - g_3 = 0. \tag{1.28}$$

This differential equation is of great importance in the applications of the Weierstrass elliptic function in physics. For a physicist it is sometimes useful to even conceive this differential equation as the definition of the Weierstrass elliptic function.

It turns out that the Weierstrass elliptic function is the general solution of the differential equation

$$\left(\frac{dy}{dz}\right)^2 = 4y^3 - g_2 y - g_3. \tag{1.29}$$

Performing the substitution $y = \wp (w)$, the Eq. (1.29) assumes the form

$$\left(\frac{dw}{dz}\right)^2 = 1,$$

which obviously has the solutions, $w = \pm z + z_0$. This implies that $y = \wp (\pm z + z_0)$ and since the Weierstrass elliptic function is even, *the general solution of Weierstrass equation* (1.29) *can be written in the form*

$$y = \wp (z + z_0). \tag{1.30}$$

In the following, we will take advantage of Weierstrass differential equation to deduce an integral formula for the inverse function of $\wp$. In order to do so, we define the function $z (y)$ as

$$z (y) := \int_y^\infty \frac{1}{\sqrt{4t^3 - g_2 t - g_3}} dt. \tag{1.31}$$

Differentiating with respect to z, one gets

$$1 = -\frac{dy}{dz} \frac{1}{\sqrt{4y^3 - g_2 y - g_3}} \Rightarrow \left(\frac{dy}{dz}\right)^2 = 4y^3 - g_2 y - g_3.$$

We just showed that the general solution of this equation is

$$y = \wp (z + z_0) .$$

Since the integral in (1.31) converges, it should vanish at the limit $y \to \infty$, i.e. $\lim_{y \to \infty} z(y) = 0$. This implies that $z = z_0$ is the position of a pole, or in other words it is congruent to $z = 0$. This means that

$$y = \wp (z + 2m\omega_1 + 2n\omega_2) = \wp (z) .$$

Substituting the above into the Eq. (1.31) yields the integral formula for Weierstrass elliptic function,

$$z = \int_{\wp(z)}^{\infty} \frac{1}{\sqrt{4t^3 - g_2 t - g_3}} dt . \tag{1.32}$$

One should wonder, how the above formula is consistent with the fact that $\wp$ is an elliptic function, and, thus, all numbers congruent to each other should be mapped to the same value of $\wp$. The answer to this question is that the integrable quantity in (1.32) has branch cuts. Depending on the selection of the path from $\wp(z)$ to infinity and more specifically depending on how many times the path encircles each branch cut, one may result in any number congruent to z or $-z$. A more precise expression of the integral formula is

$$\int_{\wp(z)}^{\infty} \frac{1}{\sqrt{4t^3 - g_2 t - g_3}} dt \sim \pm z. \tag{1.33}$$

1.2.4 The Roots of the Cubic Polynomial

We define the values of the Weierstrass elliptic function at the half-periods ω_1, ω_2 and $\omega_3 := \omega_1 + \omega_2$ as

$$e_1 := \wp (\omega_1) , \quad e_2 := \wp (\omega_3) , \quad e_3 := \wp (\omega_2) . \tag{1.34}$$

The permutation between the indices of ω's and e's is introduced for notational reasons that will become apparent later. The periodicity properties of $\wp$ combined with the fact that the latter is an even function, imply that $\wp$ is stationary at the half-periods. For example,

$$\wp' (\omega_1) = -\wp' (-\omega_1) = -\wp' (2\omega_1 - \omega_1) = -\wp' (\omega_1) ,$$

implying that $\wp' (\omega_1) = 0$. Similarly one can show that

$$\wp' (\omega_1) = \wp' (\omega_2) = \wp' (\omega_3) = 0. \tag{1.35}$$

Substituting a half-period into Weierstrass equation (1.29) yields

$$4e_i^3 - g_2 e_i - g_3 = 0. \tag{1.36}$$

The derivative of $\wp$, as shown in Eq. (1.26) has a single third order pole in each cell, congruent to $z = 0$. Thus, $\wp'$ is an elliptic function of order 3 and therefore it has exactly three roots in each cell. Since ω_1, ω_2 and ω_3 all lie within the fundamental period parallelogram, they cannot be congruent to each other, and, thus, they are these three roots and there is no other root within the fundamental period parallelogram. This also implies that ω_1, ω_2 and ω_3 are necessarily first order roots of $\wp'$. All other roots of $\wp'$ are congruent to those. Finally, the above imply that when Eq. (1.36) has a double root, the solution of the differential equation (1.29) cannot be an elliptic function.

An implication of the above is the fact that the locations $z = \omega_1$, $z = \omega_2$ and $z = \omega_3$ are the only locations within the fundamental period parallelogram, where the Laurent series of the function $\wp(z) - \wp(z_0)$ has a vanishing first order term at the region of z_0. Consequently the equation $\wp(z) = f_0$ has a double root only when f_0 equals any of the three roots e_1, e_2 or e_3. Since $\wp$ is an order two elliptic function, the complex numbers e_1, e_2 and e_3 are the only ones appearing only once in a cell, whereas all other complex numbers appear twice.

Finally, Eq. (1.36) implies that e_i are the three roots of the cubic polynomial appearing in the right hand side of Weierstrass equation, namely

$$Q(t) := 4t^3 - g_2 t - g_3 = 4(t - e_1)(t - e_2)(t - e_3). \tag{1.37}$$

This directly implies that e_i obey

$$e_1 + e_2 + e_3 = 0, \tag{1.38}$$

$$e_1 e_2 + e_2 e_3 + e_3 e_1 = -\frac{g_2}{4}, \tag{1.39}$$

$$e_1 e_2 e_3 = \frac{g_3}{4}. \tag{1.40}$$

1.2.5 Other Properties

The Weierstrass elliptic function obeys the homogeneity relation

$$\wp(z; g_2, g_3) = \mu^2 \wp\left(\mu z; \frac{g_2}{\mu^4}, \frac{g_3}{\mu^6}\right). \tag{1.41}$$

For the specific case $\mu = i$, the above relation assumes the form

$$\wp(z; g_2, g_3) = -\wp(iz; g_2, -g_3). \tag{1.42}$$

Finally, when two of the roots e_1, e_2 and e_3 coincide, the Weierstrass elliptic function degenerates to a simply periodic function. Assuming that the moduli g_2 and g_3 are real, then the existence of a double root implies that all roots are real. When the double root is larger than the simple root, the Weierstrass elliptic function assumes the form

$$\wp\left(z; 12e_0^2, -8e_0^3\right) = e_0 + \frac{3e_0}{\sinh^2\left(\sqrt{3e_0}z\right)}, \tag{1.43}$$

whereas when the double root is smaller than the simple root, it assumes the form

$$\wp\left(z; 12e_0^2, 8e_0^3\right) = -e_0 + \frac{3e_0}{\sin^2\left(\sqrt{3e_0}z\right)}. \tag{1.44}$$

If there is only one triple root, then it must be vanishing, since the three roots sum to zero. In this case, the Weierstrass elliptic function degenerates to a function that is not periodic at all, namely

$$\wp\left(z; 0, 0\right) = \frac{1}{z^2}. \tag{1.45}$$

The proofs of the homogeneity relation, as well as the double root limits of the Weierstrass elliptic function are left as an exercise for the reader.

Problems

1.1 Using the integral formula for the Weierstrass elliptic function (1.32), show that when g_2 and g_3 are real and all roots e_1, e_2 and e_3 are also real, the half-period corresponding to the largest root e_1 is congruent to a real number, whereas the half-period corresponding to the smallest root e_3 is congruent to a purely imaginary number.

Then, show that when there is one real root and two complex ones, the half-period corresponding to the real root e_2 is congruent to both a real and a purely imaginary number.

1.2 Show that at the limit when two of the roots e_1, e_2 and e_3 coincide, the Weierstrass elliptic function degenerates to a simply periodic function and can be expressed in terms of trigonometric or hyperbolic functions as described by formulae (1.43) and (1.44). Find the value of the unique period in terms of the double root. Also show that at the limit all three roots e_1, e_2 and e_3 coincide, the Weierstrass elliptic function degenerates to a non-periodic function as described by Eq. (1.45).

1.3 Use the definition (1.18) to deduce the homogeneity property of the Weierstrass elliptic function (1.41).

References

1. K. Weierstrass, *Zur Theorie der Elliptischen Funktionen*, Mathematische Werke, Bd 2 (Teubner, Berlin, 1894), pp. 245–255
2. C.G.J. Jacobi, *Fundamenta Nova Theoriae Functionum Ellipticarum* (Regiomonti, Sumtibus fratrum Borntraeger, Köönigsberg, Germany, 1829)
3. E.T. Whittaker, G.N. Watson, *A Course of Modern Analysis*, 4th edn. (Cambridge University Press, Cambridge, England, 1990). ISBN 1-438-51390-9
4. N.I. Akhiezer, *Elements of the Theory of Elliptic Functions*, Translations of Mathematical Monographs, vol. 79 (American Mathematical Society, 1990). ISBN 0-8218-4532-2
5. H. Bateman, A. Erdélyi, *Higher Transcendental Functions*, vols. 1, 2, 3 (1953). ISBN 0-486-44614-X, ISBN 0-486-44615-8, ISBN 0-486-44616-6
6. M. Abramowitz, I. Stegun, *Handbook of Mathematical Functions with Formulas, Graphs, and Mathematical Tables* (United States Department of Commerce, National Bureau of Standards (NBS), 1964)

Chapter 2
The Weierstrass Quasi-periodic Functions

Abstract In the first chapter, we used several times the fact that the derivative of an elliptic function is also an elliptic function with the same periods. However, the opposite statement is not correct; the indefinite integral of an elliptic function is not necessarily an elliptic function. This class of non-elliptic functions typically possess other interesting quasi-periodicity properties. In this chapter we study two such quasi-periodic functions, the zeta and sigma functions, which are derived from the Weierstrass elliptic function. Then, we study their basic properties and express some very useful theorems. Emphasis is given to the theorems that allow the expression of any elliptic function in terms of the aforementioned quasi-periodic functions.

The study of Weierstrass quasi-periodic functions requires familiarity with the properties of Weierstrass elliptic function, at least those studied in Chap. 1. The reader is encouraged to study the classic texts by Watson and Whittaker [1] and Akhiezer [2], for more details. The reference handbooks by Bateman and Erdélyi [3] and Abramowitz and Stegun [4] summarize many formulae on this subject. Some experience on the use of such handbooks can be proven very useful for the reader.

2.1 Quasi-periodic Weierstrass Functions

2.1.1 The Weierstrass ζ Function

The Weierstrass ζ function is defined as

$$\frac{d\zeta(z)}{dz} := -\wp(z), \tag{2.1}$$

© The Author(s), under exclusive license to Springer Nature Switzerland AG 2020
G. Pastras, *The Weierstrass Elliptic Function and Applications in Classical and Quantum Mechanics*, SpringerBriefs in Physics,
https://doi.org/10.1007/978-3-030-59385-8_2

furthermore satisfying the condition

$$\lim_{z \to 0} \left(\zeta\,(z) - \frac{1}{z} \right) := 0,$$ (2.2)

which fixes the integration constant.

Using the definition (1.18) of the Weierstrass elliptic function $\wp$, we find

$$\zeta\,(z) = \frac{1}{z}$$
$$+ \sum_{\{m,n\} \neq \{0,0\}} \left(\frac{1}{z + 2m\omega_1 + 2n\omega_2} - \frac{1}{2m\omega_1 + 2n\omega_2} + \frac{z}{(2m\omega_1 + 2n\omega_2)^2} \right).$$
(2.3)

Weierstrass ζ function is an odd function.

$$\zeta\,(-z) = -\zeta\,(z)\,.$$ (2.4)

Notice that the definition (2.1) combined with the fact that $\wp$ is an even function implies that ζ is an odd function up to a constant. The condition (2.2) fixes this constant to zero, so that ζ is an odd function.

2.1.2 Quasi-periodicity of the Function ζ

Equation (2.3) implies that in each cell defined by the periods of the corresponding $\wp$ function, the function ζ has only a first order pole with residue equal to one. As such, it cannot be an elliptic function with the same periods as $\wp$. Actually, it cannot be an elliptic function with any periods, since it is not possible to define a cell where the sum of the residues would vanish. *The Weierstrass ζ function is quasi-periodic.* Its quasi-periodicity properties can be deduced from the periodicity properties of the Weierstrass $\wp$ function. More specifically, integrating the relation $\wp\,(z + 2\omega_i) = \wp\,(z)$, we find

$$\zeta\,(z + 2\omega_i) = \zeta\,(z) + c.$$

The above relation for $z = -\omega_i$ yields $\zeta\,(\omega_i) = \zeta\,(-\omega_i) + c$. Since ζ is an odd function, the above implies that $c = 2\zeta\,(\omega_i)$, which in turn yields

$$\zeta\,(z + 2\omega_i) = \zeta\,(z) + 2\zeta\,(\omega_i)\,.$$ (2.5)

One can easily prove by induction that

$$\zeta\,(z + 2m\omega_1 + 2n\omega_2) = \zeta\,(z) + 2m\zeta\,(\omega_1) + 2n\zeta\,(\omega_2)\,.$$ (2.6)

The quantities $\zeta(\omega_1)$ and $\zeta(\omega_2)$ are related with an interesting property. Consider the contour integral

$$I = \frac{1}{2\pi i} \oint_{\partial \text{cell}} \zeta(z)\, dz.$$

Since ζ has only a first order pole with residue equal to one within a cell, Cauchy residue theorem implies that

$$I = 1.$$

Performing the contour integral along the boundary of a cell, we get

$$2\pi i I = \int_{z_0}^{z_0+2\omega_1} \zeta(z)\, dz + \int_{z_0+2\omega_1}^{z_0+2\omega_1+2\omega_2} \zeta(z)\, dz$$

$$+ \int_{z_0+2\omega_1+2\omega_2}^{z_0+2\omega_2} \zeta(z)\, dz + \int_{z_0+2\omega_2}^{z_0} \zeta(z)\, dz.$$

Shifting z by $2\omega_1$ in the second integral and by $2\omega_2$ in the third, we yield

$$2\pi i I = \int_{z_0}^{z_0+2\omega_1} (\zeta(z) - \zeta(z+2\omega_2))\, dz$$

$$- \int_{z_0}^{z_0+2\omega_2} (\zeta(z) - \zeta(z+2\omega_1))\, dz$$

$$= \int_{z_0}^{z_0+2\omega_1} (\zeta(z) - \zeta(z) - 2\zeta(\omega_2))\, dz$$

$$- \int_{z_0}^{z_0+2\omega_2} (\zeta(z) - \zeta(z) - 2\zeta(\omega_1))\, dz$$

$$= -4\omega_1 \zeta(\omega_2) + 4\omega_2 \zeta(\omega_1).$$

As a result, $\zeta(\omega_i)$ and $\zeta(\omega_i)$ are related as

$$\omega_2 \zeta(\omega_1) - \omega_1 \zeta(\omega_2) = \frac{\pi i}{2}. \tag{2.7}$$

2.1.3 The Weierstrass σ Function

Since Weierstrass elliptic function has a single second order pole in each cell, integrating it once resulted in a function (the Weierstrass ζ function) with a single first order pole in each cell. Integrating once more would lead to a logarithmic singularity in each cell. To avoid this, we define the subsequent quasi-periodic function as the exponential of the integral of the Weierstrass ζ function.

The Weierstrass σ function is defined as

$$\frac{\sigma'(z)}{\sigma(z)} := \zeta(z), \tag{2.8}$$

together with the condition

$$\lim_{z \to 0} \frac{\sigma(z)}{z} := 1, \tag{2.9}$$

which fixes the integration constant.

Integrating equation (2.3) term by term results in the following expression for the σ function.

$$\sigma(z) = z \prod_{\{m,n\} \neq \{0,0\}} \left(\frac{z + 2m\omega_1 + 2n\omega_2}{2m\omega_1 + 2n\omega_2} e^{-\frac{z}{2m\omega_1 + 2n\omega_2} + \frac{z^2}{2(2m\omega_1 + 2n\omega_2)^2}} \right). \tag{2.10}$$

This implies that σ is analytic. At the locations of the poles of $\wp$, it has first order roots.

Similarly to the definition of the ζ function, the definition (2.8) determines σ up to a multiplicative constant. This constant is set by the defining condition (2.9). The function σ is an odd function,

$$\sigma(-z) = -\sigma(z). \tag{2.11}$$

2.1.4   Quasi-periodicity of the Function σ

The Weierstrass σ function is a quasi-periodic function. Its quasi-periodicity properties follow from the corresponding properties of the Weierstrass ζ function. Integrating the Eq. (2.5) yields

$$\ln \sigma(z + 2\omega_i) = \ln \sigma(z) + 2\zeta(\omega_i) z + c,$$

or

$$\sigma(z + 2\omega_i) = C e^{2\zeta(\omega_i)z} \sigma(z),$$

where $C = e^c$. Substituting $z = -\omega_i$, we get

$$\sigma(\omega_i) = C e^{-2\zeta(\omega_i)\omega_i} \sigma(-\omega_i) = -C e^{-2\zeta(\omega_i)\omega_i} \sigma(\omega_i),$$

implying, $C = -e^{2\zeta(\omega_i)\omega_i}$. This implies that the quasi-periodicity property of the function σ is

$$\sigma(z + 2\omega_i) = -e^{2\zeta(\omega_i)(z+\omega_i)} \sigma(z). \tag{2.12}$$

Equation (2.12) can be used to prove by induction that

$$\sigma\left(z + 2m\omega_1 + 2n\omega_2\right) = (-1)^{m+n+mn} e^{(2m\zeta\left(\omega_1\right)+2n\zeta\left(\omega_2\right))(z+m\omega_1+n\omega_2)} \sigma\left(z\right). \quad (2.13)$$

2.2 Expression of Any Elliptic Function in Terms of Weierstrass Functions

A very important application of the specific construction of an elliptic function by Weierstrass is the fact that the functions $\wp$, ζ or σ can be used to express any elliptic function with the same periods. In this section, we will derive such expressions and use them in order to deduce interesting properties of a general elliptic function via its expression in terms of Weierstrass functions.

2.2.1 Expression of Any Elliptic Function in Terms of $\wp$ and $\wp'$

Let us consider an elliptic function $f(z)$. It can be trivially written as

$$f(z) = \frac{f(z) + f(-z)}{2} + \frac{f(z) - f(-z)}{2\wp'(z)} \wp'(z).$$

Both functions $\frac{f(z)+f(-z)}{2}$ and $\frac{f(z)-f(-z)}{2\wp'(z)}$ are even. Therefore, in order to express an arbitrary elliptic function in terms of $\wp$ and $\wp'$, it suffices to find an expression of an arbitrary even elliptic function in terms of $\wp$ and $\wp'$.

Assume an even elliptic function $g(z)$. As it is even, any irreducible set of poles of $g(z)$ can be divided to a set of points u_i with multiplicity r_i and another set of points congruent to $-u_i$ with equal multiplicities. In a similar manner, an irreducible set of roots of $g(z)$ can be divided to a set of points w_i with multiplicities s_i and another set of points congruent to $-w_i$ with equal multiplicities. Now consider the function

$$h(z) := g(z) \frac{\prod\limits_{i} \left(\wp(z) - \wp(u_i)\right)^{r_i}}{\prod\limits_{j} \left(\wp(z) - \wp(w_j)\right)^{s_j}}.$$

This function is obviously an elliptic function. Furthermore, it has no poles since, the poles of $g(z)$ are cancelled by the roots of the product in the numerator, the poles of the numerator are cancelled with the poles of the denominator, and the poles due to the zero's of the product in the denominator are cancelled by the zeros of $g(z)$. It follows that the function h is constant as a result of Theorem 1.2 and therefore,

$$g(z) = C \frac{\prod_j \left(\wp(z) - \wp(w_j) \right)^{s_j}}{\prod_i \left(\wp(z) - \wp(u_i) \right)^{r_i}}. \tag{2.14}$$

Summarizing, *any elliptic function can be written in terms of* $\wp(z)$ *and* $\wp'(z)$ *with the same periods. This expression is rational in* $\wp(z)$ *and linear in* $\wp'(z)$.

The above result has some interesting consequences. Consider two elliptic functions $f(z)$ and $g(z)$ with the same periods. Then, they can be both expressed as functions of $\wp$ and $\wp'$ with the same periods,

$$f(z) = F\left[\wp(z), \wp'(z) \right],$$
$$g(z) = G\left[\wp(z), \wp'(z) \right].$$

Both functions F and G are rational in their first argument and linear in the second. Bearing in mind that $\wp(z)$ and $\wp'(z)$ are also connected through the Weierstrass equation,

$$\wp'^2(z) = 4\wp^3(z) - g_2\wp(z) - g_3 = 0,$$

one can eliminate $\wp$ and $\wp'$ and result in an algebraic relation between $f(z)$ and $g(z)$. This means that *any pair of elliptic functions with the same periods are algebraically connected.* Two implications of the above sentence are:

1. *There is an algebraic relation between any elliptic function and its derivative.*
2. *There is an algebraic relation between any elliptic function and the same elliptic function with shifted argument.*

An algebraic relation between an elliptic function f and the same elliptic function with shifted argument reads

$$\sum_{k=1}^{n} f^k(z+w) \left(\sum_{l=1}^{n} c_l(w) f^l(z) \right) = 0,$$

where $c_l(w)$ are unspecified functions of w. If one interchanges z and w, they will get

$$\sum_{k=1}^{n} f^k(z+w) \left(\sum_{l=1}^{n} c_l(z) f^l(w) \right) = 0,$$

implying that the unknown functions $c_l(w)$ are necessarily powers of $f(w)$, so that the sums $\sum_{l=1}^{n} c_l(w) f^l(z)$ are symmetric polynomials in $f(z)$ and $f(w)$. This implies that the above equations assume the form of an algebraic relation between $f(z)$, $f(w)$ and $f(z+w)$, i.e. *an algebraic addition theorem.*

2.2.2 Expression of Any Elliptic Function in Terms of ζ and Its Derivatives

Consider an elliptic function $f(z)$. Assume that u_i is an irreducible set of poles of $f(z)$, which have multiplicities r_i. Furthermore, assume that the principal part of the Laurent series at the region of a pole is given by

$$f(z) \simeq \frac{c_{i,r_i}}{(z - u_i)^{r_i}} + \cdots + \frac{c_{i,2}}{(z - u_i)^2} + \frac{c_{i,1}}{z - u_i} + \mathcal{O}\left((z - u_i)^0\right).$$

Weierstrass ζ function has a single first order pole with residue equal to one in locations congruent to $z = 0$. It follows that the n-th derivative of ζ has a single $(n + 1)$-th order pole and the principal part of its Laurent series in the regime of $z = 0$ is

$$\frac{d^n \zeta(z)}{dz^n} \simeq \frac{(-1)^n n!}{z^{n+1}} + \mathcal{O}\left((z)^0\right).$$

Therefore, the function

$$g(z) := f(z) - \sum_i \left(c_{i,1} \zeta(z - u_i) - c_{i,2} \frac{d\zeta(z - u_i)}{dz} + \cdots\right.$$
$$\left. + \frac{(-1)^{r_i - 1} c_{i,r_i}}{(r_i - 1)!} \frac{d^{r_i - 1} \zeta(z - u_i)}{dz^{r_i - 1}}\right)$$

has no poles. It is not obvious though whether $g(z)$ is an elliptic function, since the function ζ is not an elliptic function. However, all derivatives of ζ are elliptic functions, and, thus,

$$g(z + 2\omega_i) - g(z) = \sum_i c_{i,1} \left(\zeta(z - u_i) - \zeta(z + 2\omega_i - u_i)\right)$$
$$= -2\zeta(\omega_i) \sum_i c_{i,1} = 0,$$

since $\sum_i c_{i,1}$ is the sum of the residues of the elliptic function $f(z)$ in a cell.

Summarizing, the function $g(z)$ is an elliptic function with no poles, and, thus, by Theorem 1.2, it is a constant function. This means that the original elliptic function $f(z)$ can be written as

$$f(z) = C + \sum_i \sum_{j=i}^{r_i} \frac{(-1)^{j-1} c_{i,j}}{(j - 1)!} \frac{d^{j-1} \zeta(z - u_i)}{dz^{j-1}}. \tag{2.15}$$

It follows that *an elliptic function is completely determined by the principal parts of its Laurent series at an irreducible set of poles, up to an additive constant.*

2.2.3 Expression of Any Elliptic Function in Terms of σ

Consider an elliptic function $f(z)$ having an irreducible set of poles u_i with multiplicities r_i and an irreducible set of roots w_j with multiplicities s_j. We recall the Theorem 1.3, which states that $\sum_i s_i \sim \sum_j r_j$. We also recall the Theorem 1.5, which states that $\sum_i s_i w_i \sim \sum_j r_j u_j$. It is always possible to select the irreducible sets of poles and roots so that $\sum_i s_i w_i = \sum_j r_j u_j$. In the following, without loss of generality, we assume that we have made such a selection.

Consider the function

$$g(z) := f(z) \, \frac{\prod\limits_i \sigma^{r_i}(z - u_i)}{\prod\limits_j \sigma^{s_j}(z - w_j)}.$$

Since the function σ has a unique first order root in each cell, which is congruent to the origin, it is obvious that $g(z)$ has neither poles nor roots.

Although it is not obvious, it turns out that the quasi-periodicity property (2.12) of the function σ implies that $g(z)$ is an elliptic function. Indeed,

$$g(z + 2\omega_k) = f(z) \, \frac{\prod\limits_i \sigma^{r_i}(z - u_i)}{\prod\limits_j \sigma^{s_j}(z - w_j)} \, \frac{\prod\limits_i \left(-e^{2\zeta(\omega_k)(z - u_i - \omega_k)}\right)^{r_i}}{\prod\limits_j \left(-e^{2\zeta(\omega_k)(z - w_j - \omega_k)}\right)^{s_j}}$$

$$= g(z) \, (-1)^{\sum\limits_i r_i - \sum\limits_j s_j} \, e^{2\zeta(\omega_k)\left[(z - \omega_k)\left(\sum\limits_i r_i - \sum\limits_j s_j\right) - \left(\sum\limits_i r_i u_i - \sum\limits_j s_j w_j\right)\right]} = g(z).$$

Since $g(z)$ is an elliptic function with no poles, it is a constant function and the original elliptic function $f(z)$ can be expressed as

$$f(z) = C \, \frac{\prod\limits_j \sigma^{s_j}(z - w_j)}{\prod\limits_i \sigma^{r_i}(z - u_i)}. \tag{2.16}$$

This implies that *an irreducible set of poles and roots of an elliptic function completely determines it, up to a multiplicative constant.*

2.3 Addition Theorems

2.3.1 Addition Theorem for $\wp$

Above, we showed that there is an algebraic addition theorem for every elliptic function. Therefore such an addition theorem exists for the Weierstrass elliptic function, too.

We define two functions of two complex variables $c_1(z, w)$ and $c_2(z, w)$ as

$$\wp'(z) = c_1(z, w)\,\wp(z) + c_2(z, w),$$
$$\wp'(w) = c_1(z, w)\,\wp(w) + c_2(z, w),$$

or in other words,

$$c_1(z, w) := \frac{\wp'(z) - \wp'(w)}{\wp(z) - \wp(w)},$$
$$c_2(z, w) := \frac{\wp(z)\,\wp'(w) - \wp(w)\,\wp'(z)}{\wp(z) - \wp(w)}.$$

We also define the function $f(x)$ of one complex variable as

$$f(x) := \wp'(x) - c_1(z, w)\,\wp(x) - c_2(z, w).$$

Clearly, the function f has only one third order pole in each cell, which is congruent to $x = 0$. Moreover, by definition it has two first order roots at $x = z$ and $x = w$. Theorem 1.3 implies that there is another first order root, which is not congruent to $x = z$ or $x = w$. Where does this root lie? Theorem 1.5 implies that the position of this root is congruent to $x = -z - w$, i.e.

$$f(-z - w) = 0.$$

The function $g(x)$ of one complex variable which is defined as,

$$g(x) := \wp'^2(x) - (c_1(z, w)\,\wp(x) + c_2(z, w))^2,$$

clearly vanishes everywhere $f(x)$ vanishes, therefore,

$$g(z) = g(w) = g(-z - w) = 0.$$

Using the Weierstrass differential equation (1.29), one can write the function $g(x)$ as,

$$g(x) = 4\wp^3(x) - c_1^2(z, w)\,\wp^2(x)$$
$$- (2c_1(z, w)\,c_2(z, w) + g_2)\,\wp(x) - \left(c_2^2(z, w) + g_3\right).$$

The fact that $g(x)$ has the three roots $x = z$, $x = w$ and $x = -z - w$, implies that the third order polynomial

$$Q(P) := 4P^3 - c_1^2(z, w) P^2$$
$$- (2c_1(z, w) c_2(z, w) + g_2) P - \left(c_2^2(z, w) + g_3\right) \quad (2.17)$$

has the roots $P = \wp(z)$, $P = \wp(w)$ and $P = \wp(-z - w)$,

$$Q(\wp(z)) = Q(\wp(w)) = Q(\wp(-z - w)) = 0.$$

In other words, the polynomial $Q(P)$ can be written as

$$Q(P) = 4(P - \wp(z))(P - \wp(w))(P - \wp(-z - w)).$$

Comparing the coefficients of the second order term of the polynomial $Q(P)$ in expression (2.17) and the above equation, we find

$$c_1^2(z, w) = 4(\wp(z) + \wp(w) + \wp(-z - w))$$

or

$$\wp(z + w) = -\wp(z) - \wp(w) + \frac{1}{4}\left(\frac{\wp'(z) - \wp'(w)}{\wp(z) - \wp(w)}\right)^2, \quad (2.18)$$

which is the desired addition theorem for the Weierstrass elliptic function. Although the expression (2.18), which is the traditional form of the addition theorem, involves the derivative of Weierstrass elliptic function, the latter can be eliminated with the use of Weierstrass differential equation (1.29) resulting in a purely algebraic relation between $\wp(z)$, $\wp(w)$ and $\wp(z + w)$.

2.3.2   Pseudo-addition Theorems for ζ and σ

The functions ζ and σ are not elliptic, and, thus, they are not guaranteed to obey algebraic addition theorems. However, the fact that any elliptic function can be written as ratio of σ functions can be used to deduce pseudo-addition theorems for them.

Consider the function $\wp(z) - \wp(w)$ as a function of z. This function, obviously has a second order pole in each cell, congruent to $z = 0$. In a trivial manner, it also has two roots congruent to $z = w$ and $z = -w$. Consequently, Eq. (2.16) implies that $\wp(z) - \wp(w)$ can be written as

$$\wp(z) - \wp(w) = A \frac{\sigma(z - w)\sigma(z + w)}{\sigma^2(z)}.$$

Writing down the principal part of the Laurent series of the above relation at the region of $z = 0$, we find

$$\frac{1}{z^2} = A \frac{\sigma(-w)\sigma(w)}{z^2},$$

implying that

$$A = -\frac{1}{\sigma^2(w)}.$$

The above results in the following pseudo-addition theorem for σ functions

$$\wp(z) - \wp(w) = -\frac{\sigma(z-w)\sigma(z+w)}{\sigma^2(z)\sigma^2(w)}. \tag{2.19}$$

Differentiating equation (2.19) with respect to z and w, we arrive at the following relations,

$$\wp'(z) = -\frac{\sigma(z-w)\sigma(z+w)}{\sigma^2(z)\sigma^2(w)}(\zeta(z-w) + \zeta(z+w) - 2\zeta(z)),$$

$$-\wp'(w) = -\frac{\sigma(z-w)\sigma(z+w)}{\sigma^2(z)\sigma^2(w)}(-\zeta(z-w) + \zeta(z+w) - 2\zeta(w)).$$

Adding them up, we find,

$$\wp'(z) - \wp'(w) = -2\frac{\sigma(z-w)\sigma(z+w)}{\sigma^2(z)\sigma^2(w)}(\zeta(z+w) - \zeta(z) - \zeta(w)),$$

which implies the following pseudo addition theorem for the ζ function

$$\frac{1}{2}\frac{\wp'(z) - \wp'(w)}{\wp(z) - \wp(w)} = \zeta(z+w) - \zeta(z) - \zeta(w). \tag{2.20}$$

Problems

2.1 Use the definitions (2.1) and (2.8) to deduce the homogeneity properties of the Weierstrass quasi-periodic functions

$$\zeta(z; g_2, g_3) = \mu\zeta\left(\mu z; \frac{g_2}{\mu^4}, \frac{g_3}{\mu^6}\right), \tag{2.21}$$

$$\sigma(z; g_2, g_3) = \frac{1}{\mu}\sigma\left(\mu z; \frac{g_2}{\mu^4}, \frac{g_3}{\mu^6}\right). \tag{2.22}$$

2.2 In problem 1.2 you showed that the Weierstrass elliptic function degenerates to simple expressions in terms of trigonometric or hyperbolic functions, when two roots coincide. Find the corresponding expressions for the functions ζ and σ. You may use the results of problem 1.2.

2.3 Prove the parity properties of Weierstrass quasi-periodic functions. Namely,

- Show that the definition (2.1) together with the defining condition (2.2) imply that ζ is an odd function.
- Show that the definition (2.8) together with the defining condition (2.9) imply that σ is an odd function.

2.4 Prove by induction the relations giving the Weierstrass quasi-periodic functions after a shift of their argument by an arbitrary period in the lattice of the corresponding Weierstrass elliptic function. Namely,

- Use Eq. (2.5) to prove (2.6).
- Use Eq. (2.12) to prove (2.13).

2.5 Use the addition theorem for Weierstrass elliptic function to show that

$$\wp (z + \omega_1) = e_1 + \frac{2e_1^2 + e_2 e_3}{\wp (z) - e_1}, \tag{2.23}$$

$$\wp (z + \omega_2) = e_3 + \frac{2e_3^2 + e_1 e_2}{\wp (z) - e_3}, \tag{2.24}$$

$$\wp (z + \omega_3) = e_2 + \frac{2e_2^2 + e_3 e_1}{\wp (z) - e_2}. \tag{2.25}$$

2.6 Use the fact that every elliptic function can be written in terms of the ζ function and its derivatives to deduce the pseudo-addition theorem for the ζ function.

2.7 Use the pseudo-addition theorem of the ζ function to deduce the addition theorem for the $\wp$ function.

2.8 Use the addition theorem of the $\wp$ function to deduce a duplication formula. Then, Use the fact that $\wp (2z)$ can be considered an elliptic function with the same periods as $\wp (z)$ to express it in terms of $\zeta (z)$ and its derivatives and result in the same duplication formula. For this purpose, you will find the results of problem 2.5 useful.

References

1. E.T. Whittaker, G.N. Watson, *A Course of Modern Analysis*, 4th edn. (Cambridge University Press, Cambridge, England, 1990). ISBN 1-438-51390-9
2. N.I. Akhiezer, *Elements of the Theory of Elliptic Functions*, Translations of Mathematical Monographs, vol. 79 (American Mathematical Society, 1990). ISBN 0-8218-4532-2
3. H. Bateman, A. Erdélyi, *Higher Transcendental Functions*, vols. 1, 2, 3 (1953). ISBN 0-486-44614-X, ISBN 0-486-44615-8, ISBN 0-486-44616-6
4. M. Abramowitz, I. Stegun, *Handbook of Mathematical Functions with Formulas, Graphs, and Mathematical Tables* (United States Department of Commerce, National Bureau of Standards (NBS), 1964)

Chapter 3
Real Solutions of Weierstrass Equation

Abstract The simplest applications of the Weierstrass functions in physics are classical mechanics problems with one degree of freedom, where the Weierstrass differential equation emerges as the conservation of energy. In such problems, the moduli are connected to physical quantities, and, thus, they are real. Furthermore, the unknown function, as well as the independent variable in the Weierstrass differential equation represent physical quantities, such as position and time, and, thus, they are real, too. It follows that we need to specify the real solutions in the real domain of the Weierstrass differential equation, in the special case that the moduli are also real. This is the task we carry out in this chapter.

The specification of the real solutions in the real domain of the Weierstrass differential equation (1.29) with real moduli g_2 and g_3 requires familiarity with the properties of the Weierstrass elliptic function, which were presented in Chap. 1. The reader may also consult the classic texts [1, 2], as well as the handbooks [3, 4].

3.1 Real Solutions of Weierstrass Equation in the Real Domain

3.1.1 The Weierstrass Elliptic Function with Real Moduli

As the general solution of equation (1.29) is given in terms of the Weierstrass elliptic function, we need to study the special properties of the latter in the special case that the moduli are real.

When the moduli g_2 and g_3 are real, there are two possible cases for the reality of the three roots:

1. All three roots are real; conventionally we define them such that $e_1 > e_2 > e_3$. In this case, *we may select the fundamental half-periods so that ω_1 is real and ω_2 is purely imaginary.* Then, they are given by the expressions,

© The Author(s), under exclusive license to Springer Nature Switzerland AG 2020
G. Pastras, *The Weierstrass Elliptic Function and Applications in Classical and Quantum Mechanics*, SpringerBriefs in Physics,
https://doi.org/10.1007/978-3-030-59385-8_3

$$\omega_1 = \int_{e_1}^{+\infty} \frac{dt}{\sqrt{4\,(t - e_1)\,(t - e_2)\,(t - e_3)}}, \tag{3.1}$$

$$\omega_2 = i \int_{-\infty}^{e_3} \frac{dt}{\sqrt{4\,(e_1 - t)\,(e_2 - t)\,(e_3 - t)}}. \tag{3.2}$$

The above expressions imply that at the limit that the two larger roots coincide, the real half-period diverges, whereas at the limit that the two smaller roots coincide, the imaginary half-period diverges.

2. There is one real root and two complex ones, which are complex conjugate to each other; conventionally, we define e_2 as the real one and we define e_1 and e_3 so that $\mathrm{Im}\,e_1 > 0$. In this case, *it is not possible to select the fundamental half-periods as in the case of the three real roots, but we may select them so that they are complex conjugate to each other*. Then, they are given by the expressions,

$$\omega_1 + \omega_2 = \int_{e_2}^{+\infty} \frac{dt}{\sqrt{4\,(t - e_1)\,(t - e_2)\,(t - e_3)}}, \tag{3.3}$$

$$\omega_1 - \omega_2 = i \int_{-\infty}^{e_2} \frac{dt}{\sqrt{4\,(t - e_1)\,(e_2 - t)\,(t - e_3)}}. \tag{3.4}$$

As the proof of this statement is a little bit technical, we kindly ask the reader to consent that it is true and continue in order to focus on its implications concerning the real solutions of the Weierstrass equation. The curious reader may study a proof of this statement in Sect. 3.2.

3.1.2 The Locus of Complex Numbers z for Whom $\wp(z)$ Is Real

We return to the investigation for real solutions of equation (1.29) with real moduli g_2 and g_3. Since the general solution of the latter is given in terms of the Weierstrass elliptic function, this investigation requires the specification of the locus of complex numbers z for whom $\wp(z)$ is real. A preliminary observation that can be made is the fact that the Weierstrass elliptic function assumes real values on the real and imaginary axes. The definition of the elliptic function $\wp$ (1.18) implies that

$$\overline{\wp(z; \omega_1, \omega_2)} = \wp(\bar{z}; \bar{\omega}_1, \bar{\omega}_2).$$

The definitions of the moduli g_2 and g_3, as given by Eqs. (1.24) and (1.25) imply that

$$g_2(\bar{\omega}_1, \bar{\omega}_2) = \overline{g_2(\omega_1, \omega_2)}, \quad g_3(\bar{\omega}_1, \bar{\omega}_2) = \overline{g_3(\omega_1, \omega_2)}$$

and consequently,

$$\overline{\wp\left(z; g_2, g_3\right)} = \wp\left(\bar{z}; \bar{g}_2, \bar{g}_3\right).$$

Thus, when the moduli g_2 and g_3 are real, it holds that

$$\overline{\wp\left(z; g_2, g_3\right)} = \wp\left(\bar{z}; g_2, g_3\right). \tag{3.5}$$

The above equation, combined with the fact that $\wp$ is even, implies that $\wp$ is real on the real and imaginary axes. Let $x \in \mathbb{R}$, then,

$$\overline{\wp\left(x; g_2, g_3\right)} = \wp\left(\bar{x}; g_2, g_3\right) = \wp\left(x; g_2, g_3\right), \tag{3.6}$$

$$\overline{\wp\left(ix; g_2, g_3\right)} = \wp\left(-ix; g_2, g_3\right) = \wp\left(ix; g_2, g_3\right). \tag{3.7}$$

But is $\wp$ real on any other points, which are not congruent to the real or imaginary axes? The answer depends on the reality of the roots e_1, e_2 and e_3. The function $\wp$ is an elliptic function of order two, and, thus, it assumes any real value (as well as any complex value) twice in every cell. The only exception to this rule are the three roots e_i, which appear only once, since they correspond to double roots of the equation $\wp\left(z\right) = e_i$. In Fig. 3.1, one cell of $\wp$ with real moduli is plotted for either three or one real root.

Dotted points are congruent to a period or a half-period. Grey dots at the boundary of the plotted cell are congruent to a black dot at another point of the boundary, and thus, they are not considered to be part of the cell. This also holds for segments that connect grey dots.

The poles are second order poles with Laurent coefficient equal to one. Therefore, as one approaches a pole from the real axis, $\wp$ tends to $+\infty$, while when one approaches a pole from the imaginary axis, $\wp$ tends to $-\infty$. Finally, as the derivative

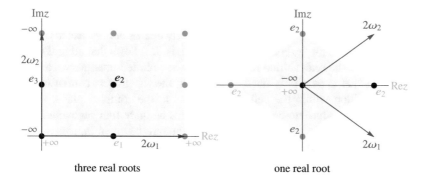

three real roots   one real root

Fig. 3.1 The values of $\wp$ on the real and imaginary axes

of $\wp$ vanishes only at the half-periods, $\wp$ is monotonous at the segments between consequent half-periods and poles.[1]

Having the above in mind, and carefully following picture 3.1, we observe that

- when there are three real roots, the function $\wp$ assumes all real values larger than e_1 twice in the segment $[0, 2\omega_1]$ on the real axis; each value appears once in $[0, \omega_1]$ and once in $[\omega_1, 2\omega_1]$. Similarly, it takes all real values smaller than e_3 twice in the segment $[0, 2\omega_2]$ on the imaginary axis.
- when there is only one real root, $\wp$ takes all real values larger than e_2 twice in the segment $[-\omega_1 - \omega_2, \omega_1 + \omega_2]$. Similarly, it takes all real values smaller than e_2 twice in the segment $[\omega_2 - \omega_1, \omega_1 - \omega_2]$ on the imaginary axis.

Therefore in the case of one real root, the function $\wp$ assumes all real values exactly twice in the cell of Fig. 3.1 at positions on the real and imaginary axes. Indeed, only e_2 appears once, as it appears only at positions congruent to each other. Consequently, the function $\wp$ cannot assume any real value at any other point within the cell, and, thus all positions where $\wp$ is real on the complex plane are congruent to a point either on the real or the imaginary axis.

In the case there are three roots, the function $\wp$ assumes all real values larger than e_1 or smaller than e_3 exactly twice in the cell at positions on the real and imaginary axes. This means that there are other positions within the cell, where $\wp$ is real and it assumes all real values between e_3 and e_1. We already know such a point, namely ω_3, where the function $\wp$ assumes the value e_2. It is a natural guess that $\wp$ is real on the horizontal and vertical axes passing through ω_3. This is indeed true. Assuming that $x \in \mathbb{R}$ and recalling that ω_1 is real, whereas ω_2 is purely imaginary,

$$\overline{\wp\,(ix + \omega_1)} = \wp\,(-ix + \omega_1) = \wp\,(ix - \omega_1)$$
$$= \wp\,(ix - \omega_1 + 2\omega_1) = \wp\,(ix + \omega_1)\,, \tag{3.8}$$

$$\overline{\wp\,(x + \omega_2)} = \wp\,(x - \omega_2) = \wp\,(x - \omega_2 + 2\omega_2) = \wp\,(x + \omega_2)\,. \tag{3.9}$$

Thus, $\wp$ assumes the values between e_2 and e_1 twice in the segment $[\omega_1, \omega_1 + 2\omega_2]$ on the shifted imaginary axis and the values between e_3 and e_2 exactly twice in the segment $[\omega_2, 2\omega_1 + \omega_2]$ on the shifted real axis. It follows that all real numbers appear twice on the real, shifted real, imaginary and shifted imaginary axes except for the roots that appear only once. As a result, the function $\wp$ cannot be real at any other position within the cell. All positions in the complex plane, where $\wp$ assumes real values have to be congruent to points on these four segments. Fig. 3.2 displays these positions as the green lattice. Although the green lattices in the two cases look identical, notice the important difference; In the case of three real roots, the

---

[1] One may think that in order to have a stationary point on the real axis, it suffices that the real part of $\wp'$ vanishes. However, this cannot be the case, as non-vanishing imaginary part of $\wp'$ would imply that $\wp$ is not real on the real axis, as we have shown. The same applies on the imaginary axis; $\wp$ is real on the imaginary axis, implying that $\wp'$ is purely imaginary. Therefore, a stationary point on the imaginary axis is necessarily a root of $\wp'$.

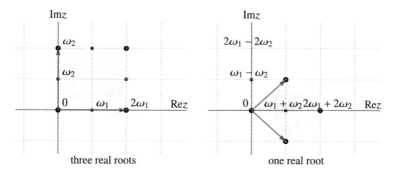

Fig. 3.2 The locus of complex numbers z for whom $\wp(z)$ is real

fundamental period parallelogram contains four cells of the "reality lattice", whereas in the case of one real root it contains only two.

Let us make a final comment; One may access all real numbers in a monotonous manner, performing a walk along the circumference of the rectangle that is defined by the origin and the three half-periods. Starting from the origin and moving along the imaginary axis towards the half-period ω_2, the Weierstrass elliptic function is an increasing function, assuming all real values smaller than e_3. Then, moving along the horizontal segment connecting ω_2 and ω_3, it monotonously increases until it reaches the value e_2. Moving along the vertical segment connecting ω_3 to ω_1, it increases until it reaches the value e_1 and finally moving back towards the origin along the real axis, it continues increasing assuming all real values larger than e_1.

3.1.3 Real Solutions of the Weierstrass Equation in the Real Domain

It is now simple to find what are the real solutions in the real domain of the equation

$$\left(\frac{dy}{dx}\right)^2 = 4y^3 - g_2 y - g_3, \tag{3.10}$$

where g_2 and g_3 are real. We know that in the complex domain, the general solution of this equation is

$$y = \wp(x - z_0), \tag{3.11}$$

where $z_0 \in \mathbb{C}$. In our problem x and y has to be real. However, z_0 is a constant of integration and has no physical meaning. It can assume any complex value, as long as y is real for any real x. This is equivalent to selecting any line on the complex plane that is parallel to the real axis. Figure 3.2 clarifies in that when there is only one real root, all such lines are congruent to the real axis itself. On the contrary, when there are three real roots, there are two options, they are congruent to the real axis, or the real axis shifted by ω_2. Consequently, *the general real solution of* (3.10) *in the real domain is*

$$y = \wp\,(x - x_0),\tag{3.12}$$

where $x_0 \in \mathbb{R}$, *when there is one real root and*

$$y = \wp\,(x - x_0)\quad\text{or}\quad y = \wp\,(x - x_0 + \omega_2),\tag{3.13}$$

where $x_0 \in \mathbb{R}$, *when there are three real roots; the appropriate choice depends on initial conditions, namely it depends on whether the initial value of* y *lies within* $[e_1, \infty)$ *or* $[e_3, e_2]$.

This behaviour of the real solutions may appear a little bizarre at first. In Sect. 4.1, this will become intuitively clear and obvious for the physicist, through a revealing example from classical mechanics.

3.2   The Half-Periods for Real Moduli

In Sect. 3.1, we specified the real solutions of the Weierstrass differential equation in the real domain for real moduli g_2 and g_3. We did so based on the fact that when the moduli are real the fundamental periods are either a real and purely imaginary one (when the associated cubic polynomial has three real roots) or complex conjugate to each other (when it has only one real root). In this section we present a proof of this statement.

In principle we would like to invert the formulae (1.24) and (1.25) in order to specify the half-periods ω_1 and ω_2 for given moduli g_2 and g_3. Although it is obvious that these formulae put certain restrictions on the fundamental periods of Weierstrass elliptic function when the moduli are real, unfortunately, the problem of inverting them is very difficult. The reader may find more on the inversion problem in the chapter on the Theta functions of [1].

For this purpose, we will apply the integral formula for Weierstrass elliptic function (1.33), for the three half-periods. In all cases, we select the integration path along the real axis.

Following Sect. 3.1, when the moduli g_2 and g_3 are real, the roots of the cubic polynomial are either all real or there is one real and two complex ones, the latter being complex conjugate to each other. In the case there are three real roots, we find

$$\omega_1 \sim \int_{e_1}^{\infty} \frac{1}{\sqrt{4\,(t - e_1)\,(t - e_2)\,(t - e_3)}}\,dt \equiv x,\tag{3.14}$$

$$\omega_2 \sim \int_{e_3}^{-\infty} \frac{1}{\sqrt{4\,(t - e_1)\,(t - e_2)\,(t - e_3)}}\,dt \equiv y,\tag{3.15}$$

whereas when there is one real and two complex roots

$$\omega_3 \sim \int_{e_2}^{\infty} \frac{1}{\sqrt{4(t-e_2)\left((z-\mathrm{Re}e_1)^2 + (\mathrm{Im}e_1)^2\right)}} dt \equiv x', \qquad (3.16)$$

$$\omega_3 \sim \int_{e_2}^{-\infty} \frac{1}{\sqrt{4(t-e_2)\left((z-\mathrm{Re}e_1)^2 + (\mathrm{Im}e_1)^2\right)}} dt \equiv y\prime. \qquad (3.17)$$

In both cases, the integrand is real everywhere in the range of integration in the first integral (the one providing x or x'), whereas it is imaginary everywhere in the range of integration in the second one (the one providing y or y'). Therefore,

$$x, x', iy, iy' \in \mathbb{R}. \qquad (3.18)$$

The above imply that when there are three real roots the half-period ω_1 is congruent to a real number and the half-period ω_2 is congruent to an purely imaginary number, i.e.

$$x = (2m_1 + 1)\omega_1 + 2n_1\omega_2, \qquad (3.19)$$
$$y = 2m_2\omega_1 + (2n_2 + 1)\omega_2, \qquad (3.20)$$

where $m_1, n_1, m_2, n_2 \in \mathbb{Z}$. On the contrary, when there is only one real root the half-period $\omega_3 = \omega_1 + \omega_2$ is congruent to both a real and a purely imaginary number, i.e.

$$x' = (2k_1 + 1)\omega_1 + (2l_1 + 1)\omega_2, \qquad (3.21)$$
$$y' = (2k_2 + 1)\omega_1 + (2l_2 + 1)\omega_2, \qquad (3.22)$$

where $k_1, l_1, k_2, l_2 \in \mathbb{Z}$.

The integrand in (3.14) never changes sign in the whole integration range. Consequently, as the upper integration limit varies from e_1 to infinity, the integral monotonically changes from 0 to its final value x. This means that x is not just a point on the real axis that is congruent to the half-period ω_1, but there is no other such position on the real axis between 0 and x. This implies that $2m_1 + 1$ and $2n_1$ are relatively prime. In a similar manner, the above statement holds for x', as well as for y and y' on the imaginary axis, and thus the pairs $\{2m_2, 2n_2 + 1\}$, $\{2k_1 + 1, 2l_1 + 1\}$ and $\{2k_2 + 1, 2l_2 + 1\}$ are also pairs of relatively prime numbers, i.e.

$$\gcd(2m_1 + 1, 2n_1) = \gcd(2m_2, 2n_2 + 1) = 1,$$
$$\gcd(2k_1 + 1, 2l_1 + 1) = \gcd(2k_2 + 1, 2l_2 + 1) = 1.$$

We may redefine the periods ω_1 and ω_2 with the use of a modular transformation, as described in Sect. 1.1,

$$\omega_1 = a\omega_1' + b\omega_2',$$
$$\omega_2 = c\omega_1' + d\omega_2',$$

where $ad - bc = 1$. Let us select $b = -2n_1$ and $d = 2m_1 + 1$ in the case of three real roots, whereas we select $b = -2l_1 - 1$ and $d = 2k_1 + 1$ in the case of one real root. Then,

$$x = (ad - bc)\,\omega_1',$$
$$y = (2m_2 a + (2n_2 + 1)\,c)\,\omega_1' + (2m_2 b + (2n_2 + 1)\,d)\,\omega_2',$$
$$x' = (ad - bc)\,\omega_1',$$
$$y' = ((2k_2 + 1)\,a + (2l_2 + 1)\,c)\,\omega_1' + ((2k_2 + 1)\,b + (2l_2 + 1)\,d)\,\omega_2'.$$

We managed to eliminate ω_2' in the expressions for x and x'. However, we should ask whether there exist modular transformations for the specific selections of b and d made above. In other words, are there integer solutions for a and c to the equation

$$ad - bc = 1 \tag{3.23}$$

for the specific selections of b and d made above? This equation is a linear Diophantine equation of the form $\alpha x + \beta y = \gamma$ and it is known that such equations have integer solutions, as long as γ is a multiple of the greatest common divisor of α and β. In both cases b and d are relatively prime, and, thus, their greatest common divisor is equal to one. Therefore, in both cases, Eq. (3.23) does have solutions. The reader may find more on linear Diophantine equations in any introductory number theory textbook, e.g. [5]. Furthermore, the parity of the selected b and d implies that in the case of three real roots, a will be odd, while in the case of one real root a and c have to be of opposite parity. The above imply that

$$x = \omega_1', \tag{3.24}$$
$$y = m_2'\omega_1' + \left(2n_2' + 1\right)\omega_2', \tag{3.25}$$
$$x' = \omega_1', \tag{3.26}$$
$$y' = \left(2k_2' + 1\right)\omega_1' + 2l_2'\omega_2', \tag{3.27}$$

where $m_2', n_2', k_2', l_2' \in \mathbb{Z}$. In both cases, the new lattice has been formed so that the fundamental period parallelogram has one side parallel to the real axis.

Let us now focus in the case of three real roots. As we commented above, the Weierstrass function ranges between $+\infty$ and e_1 between the origin and the real half-period x. The fact that $\wp$ is even means that it actually ranges between $+\infty$ and e_1 in the whole period between $-x$ and x and consequently in the whole real axis. The periodicity properties of the elliptic functions imply that this holds in any shifted axis by any multiple of $2\omega_2$. If $|2n_2' + 1| > 2$, then the segment of the imaginary axis from 0 to y crosses such a line, and, thus, there is point on this segment where $\wp$ takes a value larger or at most equal to e_1 (see Fig. 3.3).

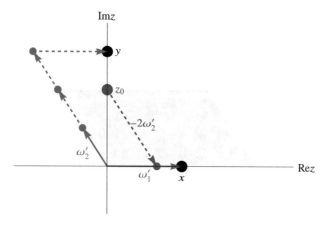

Fig. 3.3 An example of the inconsistency that appears whenever $n_2' \neq 0$. In this example it is assumed that $m_2' = n_2' = 1$. It is evident that z_0 is congruent to a point on the real axis, thus, $\wp(z_0) \in [e_1, +\infty)$. At the same time z_0 lies in the imaginary axis between 0 and y, therefore $\wp(z_0) \in (-\infty, e_3]$. Obviously, since $e_1 > e_3$, these statements cannot be both true

But we have already stated that Weierstrass function is real on the imaginary axis and changes monotonically from $-\infty$ at $z = 0$ to e_3 at $z = y$. Since $e_3 < e_1$, This is not possible. Therefore the only consistent possibility is $n_2' = 0$ (or $n_2' = -1$) and

$$x = \omega_1',$$
$$y = m_2'\omega_1' + \omega_2'.$$

A further modular transformation of the form

$$\omega_1' = \omega_1'',$$
$$\omega_2' = m_2'\omega_1'' - \omega_2'',$$

results in

$$x = \omega_1'', \tag{3.28}$$
$$y = \omega_2'', \tag{3.29}$$

meaning that when there are three real roots, the two fundamental half-periods can be selected so that one of them is real and the other purely imaginary. If such a selection is performed, then their values are equal to x and y, which are given by the integral formulae (3.14) and (3.15). The case $n_2' = -1$ simply results in $y = -\omega_2''$.

In a similar manner, in the case of one real root, monotonicity of the Weierstrass elliptic function on the imaginary axis between 0 and y' implies that $l_2' = 1$ (or $l_2' = -1$) and

$$x' = \omega_1',$$
$$y' = \left(2k_2' + 1\right)\omega_1' + 2\omega_2'.$$

In this case, it is easy to show that there is no modular transformation that would preserve the reality of ω_1 and simultaneously set ω_2 to an imaginary value. Such a transformation would necessarily be of the form

$$\omega_1' = \omega_1'',$$
$$\omega_2' = c\omega_1'' + \omega_2''$$

and it would transform the half-periods to the form,

$$x' = \omega_1'',$$
$$y' = \left(2k_2' + 2c + 1\right)\omega_1'' + 2\omega_2''.$$

Clearly, the coefficient of ω_1'' cannot be set to zero by such a transformation. On the contrary, in this case, one may perform the transformation

$$\omega_1' = \omega_1'' + \omega_2'',$$
$$\omega_2' = -\left(k_2' + 1\right)\omega_1'' - k_2'\omega_2'',$$

to find

$$x' = \omega_1'' + \omega_2'', \tag{3.30}$$
$$y' = -\omega_1'' + \omega_2'', \tag{3.31}$$

meaning that ω_1 and ω_2 can be defined so that they are complex conjugate to each other. If such a selection is performed, then the fundamental half-periods are given by $\left(x' \pm y'\right)/2$, where x' and y' are given by the integral formulae (3.16) and (3.17) respectively.

References

1. E.T. Whittaker, G.N. Watson, *A Course of Modern Analysis*, 4th edn. (Cambridge University Press, Cambridge, England, 1990). ISBN 1-438-51390-9
2. N.I. Akhiezer, *Elements of the Theory of Elliptic Functions*, Translations of Mathematical Monographs, vol. 79 (American Mathematical Society, 1990). ISBN 0-8218-4532-2
3. H. Bateman, A. Erdélyi, *Higher Transcendental Functions*, vols. 1, 2, 3 (1953). ISBN 0-486-44614-X, ISBN 0-486-44615-8, ISBN 0-486-44616-6
4. M. Abramowitz, I. Stegun, *Handbook of Mathematical Functions with Formulas, Graphs, and Mathematical Tables* (United States Department of Commerce, National Bureau of Standards (NBS), 1964)
5. G.E. Andrews, *Number Theory* (Dover Publications Inc., 2000). ISBN 978-0486682525

Chapter 4
Applications in Classical Mechanics

Abstract In this chapter we study applications of Weierstrass functions in classical mechanics, particularly in one-dimensional problems. The common characteristic of these problems is the fact that the Weierstrass differential equation emerges as the conservation of energy. The benefits of this study are going to be twofold: Firstly, we are going to obtain analytic solutions to some basic mechanics problems, such as the cubic potential or the simple pendulum, which allow the comprehension of the qualitative behaviour of these systems. Secondly, we are going to acquire a better physical understanding of the properties of the Weierstrass elliptic function, especially those analysed in Chap. 3, through the conception of the latter as the solution to a mechanical problem.

In the first two chapters we performed a fast introduction to the theory of Weierstrass elliptic and related functions. In the third chapter we focused on properties of the latter that are going to be crucial for applications in physics. Although the first three chapters cover the necessary background material for this chapter, the reader may find the handbooks [1, 2], which provide fast access to formulae, useful.

4.1 Point Particle in a Cubic Potential

4.1.1 Problem Definition

Firstly, we study a problem where the Weierstrass differential equation emerges directly as the conservation of energy. We consider a point particle moving in one dimension under the influence of a force that is a quadratic function of position. Without loss of generality, we select the origin of the coordinate system as the position of extremal force and we select units, such that the mass of the particle equals 2 and the coefficient of the quadratic term of the force equals 12. With this conventions, the equation of motion is written as

$$2\ddot{x} = F_0 + 12x^2. \tag{4.1}$$

This equation can be integrated once to the form of conservation of energy. Fixing the integration constant so that the potential vanishes at $x = 0$ we get

$$\dot{x}^2 + V(x) = E, \quad V(x) = -F_0 x - 4x^3 \tag{4.2}$$

or

$$\dot{x}^2 = 4x^3 + F_0 x + E. \tag{4.3}$$

Obviously, there is no local minimum of the potential when $F_0 > 0$. In this case all motions of the problem are scattering solutions evolving from $+\infty$ to a minimum value of x and then back to $+\infty$. On the contrary, as shown in Fig. 4.1, when $F_0 < 0$ there is a local maximum of the potential at $x = x_0 \equiv \sqrt{-F_0/12}$ and a local minimum at $x = -x_0$, and, thus, a range of values for the energy E, namely,

$$|E| < E_0, \quad E_0 = \left(-\frac{F_0}{3}\right)^{\frac{3}{2}}, \tag{4.4}$$

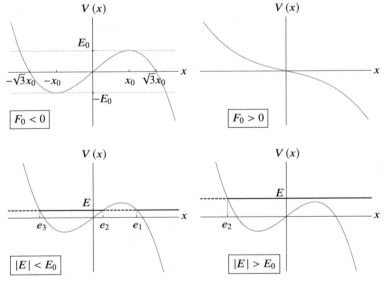

Fig. 4.1 The cubic potential for $F_0 < 0$ (top left) and $F_0 > 0$ top right. In the case $F_0 < 0$, there may exist one (bottom right) or two (bottom left) possible motions depending on the value of the energy

for whom the equation $V(x) = E$ has three roots and consequently there are two possible kinds of motion: One of them is a scattering solution evolving from $+\infty$ to a minimum value of x and then back to $+\infty$ and the other is an oscillating solution in the region of the local minimum of the potential.

4.1.2 Problem Solution

In the language of classical mechanics, it becomes obvious why the Weierstrass differential equation has two independent real solutions when there are three real roots and only one when there is one real root, as we showed in Chap. 3. The roots play the role of the extrema of the motion, which are indeed the positions where the velocity vanishes, as required for the roots. Furthermore, the solution that always exists is the one defined on the real axis,

$$x = \wp\,(t - t_0; -F_0, -E)\,, \tag{4.5}$$

which contains the pole, and, thus, it corresponds to the scattering solution. The solution on the shifted real axis,

$$x = \wp\,(t - t_0 + \omega_2; -F_0, -E,)\,, \tag{4.6}$$

is bounded between e_3 and e_2 and corresponds to the oscillating solution in the region of the local minimum. Finally, we would like to commend that from the point of view of classical mechanics, it is natural that the Weierstrass elliptic function (or more literally the solution of the Weierstrass equation) has order equal to two. In every position there are two possible initial conditions that correspond to the same energy, depending on the direction of the initial velocity. This is mirrored to the fact that the same real value appears in two non-congruent positions in every cell. This does not apply only at the extrema of motion, where the appropriate initial velocity vanishes and indeed these correspond to the roots of the Weierstrass function, which appear only once in every cell.

Finally, let's make an interesting observation concerning the "time of flight" for scattering solutions and the period of the oscillating solutions. The former is the distance between two consecutive poles on the real axis, which obviously equals

$$T_{\text{scattering}} = 2\omega_1\,. \tag{4.7}$$

Similarly, the period of the oscillating solutions is the distance between two consecutive appearances of the same root. This is also clearly

$$T_{\text{oscillating}} = 2\omega_1\,. \tag{4.8}$$

Therefore, *for the energies that an oscillating solution exists, the "time of flight" of the scattering solution and the period of the oscillating solution with the same energy are equal.*

4.1.3 The Role of the Imaginary Period

The Weierstrass elliptic function naturally describes the motion of a point particle in a cubic potential. The real period of the Weierstrass function ($2\omega_1$ in the case of three real roots, $2\omega_3$ in the case of one real root) is equal to the "time of flight" or the period of the motion. Is there any physical meaning for the imaginary period?

It is easy to answer this question using the homogeneity transformation (1.41). This relation with the specific selection $\mu = i$ implies that

$$\wp\left(iz; g_2, g_3\right) = -\wp\left(z; g_2, -g_3\right). \tag{4.9}$$

It is a direct consequence that $\wp\left(iz; g_2, g_3\right)$ obeys the differential equation,

$$\left(\frac{d\wp\left(iz; g_2, -g_3\right)}{dz}\right)^2 = -4\wp^3\left(iz; g_2, -g_3\right) + g_2\wp\left(iz; g_2, -g_3\right) - g_3.$$

Selecting $g_2 = -F_0$ and $g_3 = E$, we find that the function $\wp\left(iz; -F_0, -E\right)$ obeys the differential equation,

$$\left(\frac{d\wp\left(iz; -F_0, -E\right)}{dz}\right)^2 = -4\wp^3\left(iz; -F_0, -E\right) - F_0\wp\left(iz; -F_0, -E\right) - E,$$

and, thus, it is a solution to another one-dimensional point particle problem, namely,

$$\dot{y}^2 + \tilde{V}\left(y\right) = -E, \quad \tilde{V}\left(y\right) = y^3 + F_0 y = -V\left(y\right). \tag{4.10}$$

This is clearly the problem of motion of a point particle in the inverted potential to that of the initial problem, having the opposite energy.

In this problem, the point particle is moving in the complementary region than that of the initial problem. This is clearly depicted in Fig. 4.2. The unbounded motion of the point particle under the influence of the potential $\tilde{V}$ is given by

$$y = \wp\left(i(t - t_0); -F_0, -E\right), \tag{4.11}$$

while, when there is a bounded one, it is given by

$$y = \wp\left(i(t - t_0) + \omega_1; -F_0, -E\right). \tag{4.12}$$

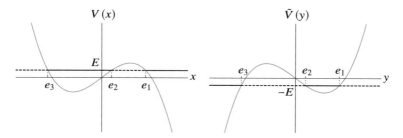

Fig. 4.2 The original and inverted point particle problems

The physical importance of the imaginary period is now obvious. The "time of flight" of the unbounded motion, as well as the period of the bounded motion in this inverted problem are given by the imaginary period of $\wp\,(z; -F_0, -E,\,)$, i.e.

$$\tilde{T}_{\text{scattering}} = \tilde{T}_{\text{oscillating}} = -2i\omega_2. \tag{4.13}$$

4.2 The Simple Pendulum

The Weierstrass elliptic function naturally describes a point particle in a cubic potential, due to the fact that the conservation of energy takes the form of the Weierstrass equation (1.29). Its applications though are not limited to this problem. There are several other important problems with one degree of freedom that can be transformed to that of a cubic potential with an appropriate coordinate transformation.

4.2.1 Problem Definition and Equivalence to Weierstrass Equation

One simple and important problem that can be transformed to a cubic potential problem is the simple pendulum. The equation of motion reads

$$\ddot{\theta} = -\omega^2 \sin\theta. \tag{4.14}$$

It can be integrated once to take the form of energy conservation,

$$\frac{1}{2}\dot{\theta}^2 + V\,(\theta) = E, \quad V\,(\theta) = -\omega^2 \cos\theta. \tag{4.15}$$

Fig. 4.3 The simple
pendulum potential

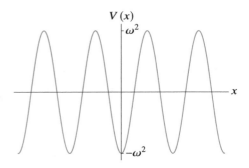

The potential $V(\theta)$ is plotted in Fig. 4.3. The form of the potential indicates that

- there are no solutions for $E < -\omega^2$,
- there are oscillating solutions for $|E| < \omega^2$,
- there are continuously rotating solutions for $E > \omega^2$.

We perform the change of variable

$$- \omega^2 \cos \theta = 2y + \frac{E}{3}. \tag{4.16}$$

Then, the conservation of energy assumes the form

$$\dot{y}^2 = 4y^3 - \left(\frac{E^2}{3} + \omega^4\right) y - \frac{E}{3}\left(\left(\frac{E}{3}\right)^2 - \omega^4\right). \tag{4.17}$$

This is the standard form of Weierstrass differential equation. The solution for y of course should be real, but we should also ensure that

$$\left| 2y + \frac{E}{3} \right| < \omega^2, \tag{4.18}$$

so that the change of variable (4.16) leads to a real θ.

4.2.2 Problem Solution and Classification of Solutions

The roots of the cubic polynomial in the right hand side of Eq. (4.17) turn out to acquire simple expressions. After some algebra, this polynomial assumes the form

$$Q(y) = 4\left(y - \frac{E}{3}\right)\left(y + \frac{E}{6} - \frac{\omega^2}{2}\right)\left(y + \frac{E}{6} + \frac{\omega^2}{2}\right). \tag{4.19}$$

Thus, the three roots are all real for any value of the energy constant E. They equal

$$x_1 := \frac{E}{3}, \quad x_2 := -\frac{E}{6} + \frac{\omega^2}{2}, \quad x_3 := -\frac{E}{6} - \frac{\omega^2}{2}. \qquad (4.20)$$

We use the notation x_i for the roots as given by Eqs. (4.20) to reserve the notation e_i for the roots appropriately ordered. As there are always three real roots, there are always two independent real solutions of Eq. (4.17). They are

$$y = \wp(t - t_0; g_2(E), g_3(E)), \qquad (4.21)$$
$$y = \wp(t - t_0 + \omega_2; g_2(E), g_3(E)), \qquad (4.22)$$

where

$$g_2(E) = \frac{E^2}{3} + \omega^4, \quad g_3 = \frac{E}{3}\left(\left(\frac{E}{3}\right)^2 - \omega^4\right). \qquad (4.23)$$

The ordering of the three roots x_i depends on the value of the energy constant E, as shown in Fig. 4.4. As the roots e_i are defined so that $e_1 > e_2 > e_3$, the identification of x_i with e_i depends on the value of the constant E. The appropriate assignments are summarized in Table 4.1.

The unbounded solution (4.21) ranges from e_1 to infinity, whereas the bounded one (4.22) ranges between e_3 and e_2. Using the appropriate assignment of roots for each energy region, we find that the range of $-\omega^2 \cos\theta = 2y + E/3 = 2y + x_1$, depending on the value of the energy, is given in Table 4.2.

The Table 4.2 clearly implies that the unbounded solution never corresponds to a real θ. The bounded solution corresponds to a real solution only when $E > -\omega^2$.

Fig. 4.4 The roots of the cubic polynomial (4.17) as function of the energy constant E

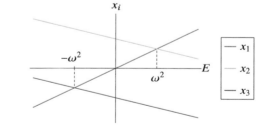

Table 4.1 The orderings of the roots x_i depending of the value of the energy

	Ordering of roots
$E < -\omega^2$	$e_1 = x_2, e_2 = x_3, e_3 = x_1$
$\|E\| < \omega^2$	$e_1 = x_2, e_2 = x_1, e_3 = x_3$
$E > \omega^2$	$e_1 = x_1, e_2 = x_2, e_3 = x_3$

Table 4.2 The range of $-\omega^2 \cos\theta$ depending of the value of the energy

	The range of $-\omega^2 \cos\theta$	
	For the unbounded solution	For the bounded solution
$E < -\omega^2$	$[\omega^2, +\infty)$	$[E, -\omega^2]$
$|E| < \omega^2$	$[\omega^2, +\infty)$	$[-\omega^2, E]$
$E > \omega^2$	$[E, +\infty)$	$[-\omega^2, \omega^2]$

Thus, as expected by the form of the potential, the pendulum problem has a real solution only when $E > -\omega^2$ and this is given by the bounded solution (4.22).

Since the solution is single valued for $\cos\theta$, in order to find an analytic expression for θ, one has to match appropriate patches, so that the overall solution is everywhere continuous and smooth. It is not difficult to show that selecting initial conditions, so that $\theta(0) = 0$ and $\dot\theta(0) > 0$, the appropriate expression for the angle theta is

$$\theta = \begin{cases} (-1)^{\left\lfloor \frac{t}{2\omega_1} \right\rfloor} \arccos\left[\frac{1}{\omega^2}\left(2\wp(t + \omega_2) + \frac{E}{3}\right)\right], & E < \omega^2, \\ (-1)^{\left\lfloor \frac{t}{\omega_1} \right\rfloor} \arccos\left[\frac{1}{\omega^2}\left(2\wp(t + \omega_2) + \frac{E}{3}\right)\right] + 2\pi \left\lfloor \frac{t}{2\omega_1} + \frac{1}{2} \right\rfloor, & E > \omega^2. \end{cases}$$
$$(4.24)$$

It is left as an exercise for the reader to verify that the above expressions are everywhere continuous and smooth.

There is naturally a qualitative change of the form of the solutions at $E = \omega^2$, which is mirrored in the ordering of the roots. The period of the oscillatory motions $(E < \omega^2)$ is

$$T_{\text{oscillating}} = 4\omega_1. \qquad (4.25)$$

while the period (or more literally the quasi-period) of the rotating motions $(E > \omega^2)$ is

$$T_{\text{rotating}} = 2\omega_1. \qquad (4.26)$$

The time evolution of θ is sketched in Table 4.3. The difference between the two expressions is due to the change of the topology of the solution. A half-period of the solution corresponds to the transition from the equilibrium position to the maximum displacement position. In the case of oscillatory motion four such segments are required to complete a period, as after two segments the pendulum is back at the

Table 4.3 The elliptic solutions of the simple pendulum equation at multiples of the half-period

	$\theta(0)$	$\theta(\omega)$	$\theta(2\omega)$	$\theta(3\omega)$	$\theta(4\omega)$
$E < -\omega^2$	–				
$|E| < \omega^2$	0	$\left(\pi - \arccos\frac{E}{\omega^2}\right)$	0	$-\left(\pi - \arccos\frac{E}{\omega^2}\right)$	0
$E > \omega^2$	0	π	2π	3π	4π

equilibrium position but with inverted velocity. On the contrary, in the case of a rotating solution, the maximum displacement equals π, the velocity is never inverted, and, thus, only two half-periods are required to complete a period.

4.3 Point Particle in a Hyperbolic Potential

4.3.1 Problems Definition and Equivalence to Weierstrass Equation

Now let's consider the case of a point particle moving in one dimension under the influence of a hyperbolic force. There are four such possible cases, namely,

$$\ddot{x} = \omega^2 \sinh x, \tag{4.27}$$

$$\ddot{x} = -\omega^2 \sinh x, \tag{4.28}$$

$$\ddot{x} = \omega^2 \cosh x, \tag{4.29}$$

$$\ddot{x} = -\omega^2 \cosh x. \tag{4.30}$$

We will study all those four cases simultaneously by writing the equation of motion as

$$\ddot{x} = -s\omega^2 \frac{e^x + te^{-x}}{2}. \tag{4.31}$$

The parameters s and t take the values ± 1. Appropriate selection of s and t results in any of the four possible hyperbolic forces. Equation (4.31) can be integrated once to yield

$$\frac{1}{2}\dot{x}^2 + V(x) = E, \quad V(x) = s\frac{\omega^2}{2}\left(e^x - te^{-x}\right). \tag{4.32}$$

The potential energy is plotted in Fig. 4.5 for all the four cases that we are studying. Considering the form of the potential, we obtain a qualitative picture for the behaviour of the solutions. In the case $s = +1$ and $t = -1$, we expect to find oscillating solutions with energy $E > m^2$ and no solutions for $E < m^2$. In the case $s = -1$ and $t = -1$, we expect to find two different classes of solutions; for $E < -m^2$ we expect to find reflecting scattering solutions since the point particle does not have enough energy to overcome the potential barrier, whereas for $E > -m^2$, we expect to find transmitting scattering solutions since the particle overcomes the potential barrier. Finally, in the case $t = +1$, we expect to find reflecting scattering solutions for all energies.

Performing the change of variable

$$-s\frac{\omega^2}{2}e^x = 2y - \frac{E}{3}, \tag{4.33}$$

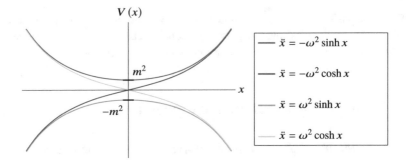

Fig. 4.5 The potential for the four cases of a hyperbolic force

equation (4.32) takes the standard Weierstrass form

$$\dot{y}^2 = 4y^3 - \left(\frac{1}{3}E^2 + t\frac{\omega^4}{4}\right)y + \frac{E}{3}\left(\frac{1}{9}E^2 + t\frac{\omega^4}{8}\right). \tag{4.34}$$

The change of variable (4.33) transforms the problem of the motion of a particle under the influence of a hyperbolic force to yet another one-dimensional problem, describing the motion of a particle with zero energy under the influence of a cubic potential, which has already been studied in Sect. 4.1. The form of (4.33) implies that real solutions of this equation correspond to real values of the initial variable x only when $2y - \frac{E}{3}$ has the same sign as s.

4.3.2 Four Problems Solved by the Same Expression

It is interesting to understand how the same equation can be used to describe a variety of solutions that exhibit qualitatively different behaviour, as suggested by the form of the potential in the four cases of hyperbolic forces under study.

Equation (4.34) is of the standard Weierstrass form (1.29) with a specific selection for the constants g_2 and g_3. Equation (4.34) is solved by

$$y = \wp(t; g_2(E, t), g_3(E, t)), \tag{4.35}$$

$$y = \wp(t + \omega_2; g_2(E, t), g_3(E, t)), \tag{4.36}$$

bearing in mind that the second solution is valid only when there are three real roots. The coefficients g_2 and g_3 are given by

$$g_2(E, t) = \frac{1}{3}E^2 + t\frac{\omega^4}{4}, \quad g_3(E, t) = -\frac{E}{3}\left(\frac{1}{9}E^2 + t\frac{\omega^4}{8}\right) \tag{4.37}$$

and the related cubic polynomial is

$$Q(x) = 4x^3 - \left(\frac{1}{3}E^2 + t\frac{\omega^4}{4}\right)x + \frac{E}{3}\left(\frac{1}{9}E^2 + t\frac{\omega^4}{8}\right). \tag{4.38}$$

The roots of the cubic polynomial can be easily obtained noting that $x = E/6$ is one of them. The three roots of $Q(x)$ are

$$x_1 = \frac{E}{6}, \quad x_{2,3} = -\frac{E}{12} \pm \frac{1}{4}\sqrt{E^2 + t\omega^4}. \tag{4.39}$$

As in the case of the pendulum, we use the notation x_i for the roots of $Q(x)$ as written in Eqs. (4.39) and reserve the notation e_i for the ordered roots of $Q(x)$. The roots x_i are plotted as functions of the energy constant E in Fig. 4.6.

The Weierstrass function allows for a unifying description of the elliptic solutions of both sinh and cosh forces. Different classes of solutions simply correspond to different orderings of the roots x_i. Figure 4.6 suggests that there are four distinct cases for the ordering of the three roots x_i, which are summarized in Table 4.4.

The unbounded solution ranges from e_1 to infinity when there are three real roots and from e_2 to infinity when there is only one real root, whereas the bounded solution ranges from e_3 to e_2. Using Eq. (4.33), we can specify the range of $-s\frac{\omega^2}{2}e^x$ in all cases. The results are summarized in Table 4.5. In all cases, the sign of $-s\frac{\omega^2}{2}e^x$ does not alternate within its range. Consequently, each solution corresponds to a real solution for exactly one value of the sign s.

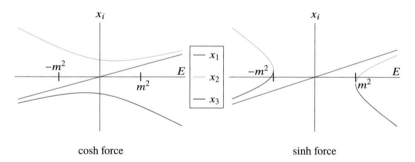

cosh force   sinh force

Fig. 4.6 The roots of the cubic polynomial (4.38) as function of the energy constant E

Table 4.4 The possible orderings of the roots x_i

	Reality of roots	Ordering of roots
$t = +1$	3 real roots	$e_1 = x_2, e_2 = x_1, e_3 = x_3$
$t = -1, E > \omega^2$	3 real roots	$e_1 = x_1, e_2 = x_2, e_3 = x_3$
$t = -1, |E| < \omega^2$	1 real, 2 complex roots	$e_1 = x_2, e_2 = x_1, e_3 = x_3$
$t = -1, E < -\omega^2$	3 real roots	$e_1 = x_2, e_2 = x_3, e_3 = x_1$

Table 4.5 The solutions for $-s\frac{\omega^2}{2}e^x$ and their range

	$-s\frac{\omega^2}{2}e^x$	Unbounded range	Bounded range
$t=+1$	$2(y-e_2)$	$[2(e_1-e_2),+\infty)$	$[-2(e_2-e_3),0]$
$t=-1, E>\omega^2$	$2(y-e_1)$	$[0,+\infty)$	$[-2(e_1-e_3),-2(e_1-e_2)]$
$t=-1,$ $|E|<\omega^2$	$2(y-e_2)$	$[0,+\infty)$	–
$t=-1,$ $E<-\omega^2$	$2(y-e_3)$	$[2(e_1-e_3),+\infty)$	$[0,2(e_2-e_3)]$

Table 4.6 The extrema of the motion

	Unbounded			Bounded		
	$x(0)$	$x(\Omega)$	$x(2\Omega)$	$x(0)$	$x(\Omega)$	$x(2\Omega)$
	$s=-1$					
$t=+1$	$+\infty$	$\ln\frac{4(e_1-e_2)}{\omega^2}$	$+\infty$	–		
$t=-1,$ $E>\omega^2$	$+\infty$	$-\infty$	$+\infty$	–		
$t=-1,$ $|E|<\omega^2$	$+\infty$	$-\infty$	$+\infty$	–		
$t=-1,$ $E<-\omega^2$	$+\infty$	$\ln\frac{4(e_1-e_3)}{\omega^2}$	$+\infty$	$-\infty$	$\ln\frac{4(e_2-e_3)}{m^2}$	$-\infty$
	$s=+1$					
$t=+1$	–			$\ln\frac{4(e_2-e_3)}{\omega^2}$	$-\infty$	$\ln\frac{4(e_2-e_3)}{\omega^2}$
$t=-1,$ $E>\omega^2$	–			$\ln\frac{4(e_1-e_2)}{\omega^2}$	$\ln\frac{4(e_1-e_3)}{\omega^2}$	$\ln\frac{4(e_1-e_2)}{\omega^2}$
$t=-1,$ $|E|<\omega^2$	–			–		
$t=-1,$ $E<-\omega^2$	–			–		

In Table 4.6, we sketch the time evolution of each solution. In this table 2Ω stands for the real period of y, which is equal to $2\omega_1$ when there are three real roots and $2(\omega_1+\omega_2)$ when there is only one real root.

All solutions take the following form

$$x(t) = \ln\left[-s\frac{2}{\omega^2}\left(2\wp(t+z_0; g_2(E), g_3(E)) - \frac{E}{3}\right)\right], \tag{4.40}$$

for all choices of the overall sign s and an appropriate choice of the complex integration constant z_0. In particular:

- In the case $s=-1$ and $t=+1$, as expected from the form of the potential, only reflecting solutions, coming from and going to the right are found for all energies. In this case there are always three real roots and the solution is given by the unbounded solution. If we select initial conditions, so that the particle is at the minimum position at $t = t_0$, we need to select $z_0 = \omega_1 - t_0$. The "time of flight" equals $T = 2\omega_1$.

- In the case $s = +1$ and $t = +1$, as expected from the form of the potential, only reflecting solutions, coming from and going to the left exist for all energies. In this case there are always three real roots and the solution is given by the bounded solution. Selecting initial conditions, so that the particle is at the maximum position at $t = t_0$, we have to select $z_0 = \omega_2 - t_0$. The "time of flight" equals $T = 2\omega_1$.
- In the case $s = -1$ and $t = -1$, the form of the potential suggests that there are two possible cases depending on the energy.

  - When $E < -\omega^2$, the particle does not have enough energy to overcome the potential barrier. Therefore, there are reflecting solutions coming from and going to either of the two directions. In this case there are three real roots. The particles coming from the right are described from the unbounded solution, while the particles coming from the left are described from the bounded solution. Selecting initial condition, so that the particle is at the extremal position at $t = t_0$ requires the selection $z_0 = \omega_1 - t_0$ for the particles coming from the right and $z_0 = \omega_1 + \omega_2 - t_0$ for the particles coming from the left. The "time of flight" in both cases equals $T = 2\omega_1$.
  - When $E > -\omega^2$, the particle has enough energy to overcome the potential barrier. Therefore, there are two classes of transmitting solutions coming from either direction. It is interesting though that the reality of the three roots depends on whether the energy is smaller or larger than the critical value ω^2.

    · When $E < \omega^2$, there is only one real root and both left-incoming and right-incoming particles are described by the unbounded solution. The "time of flight" in both cases equals $T = \omega_3$. Selecting initial conditions such that the particle at $t = t_0$ lies at the origin, one should select $z_0 = \omega_3/2 - t_0$ for a particle coming from the right and $z_0 = 3\omega_3/2 - t_0$ for a particle coming from the left.
    · When $E > \omega^2$, there are three real roots. Apart from this, the situation is similar to the case $E < \omega^2$, with the substitution of ω_3 with ω_1.

- Finally, when $s = +1$ and $t = -1$, there are oscillatory solutions only when $E > \omega^2$, as expected by the form of the potential. The period of the oscillation equals $T = 2\omega_1$. Selecting initial conditions so that the particle lies at its minimum position at $t = t_0$ yields $z_0 = -t_0$.

Problems

4.1 Find the energies for which the solution for the cubic potential degenerates to a Weierstrass function with a double root, thus a simply periodic function. Find the special expressions and the periods of the motion at these limits. Verify that in one of these limits, the unique period is equal to the period of small oscillations in the region of the local minimum.

4.2 In Sect. 4.1, we showed that while the real period of the Weierstrass elliptic function equals the "time of flight" or the period of the motion of a point particle with energy E in a cubic potential, the absolute value of the imaginary period equals the corresponding quantities for the motion of a point particle with energy $-E$ in the inverted potential. However, the inverted potential is identical to the initial one if a coordinate reflection is performed. Thus, the absolute value of the imaginary period should also equal the "time of flight" or period of the motion of a point particle with energy $-E$ in the initial potential. Verify this using expressions (3.1), (3.2), (3.3) and (3.4) .

4.3 In Sect. 4.1, we showed that when a bounded oscillatory motion exists in a cubic potential, its period equals the "time of flight" of the scattering solution for the same energy. This happens due to a discrete symmetry of the conservation of energy equation that exists when there are three real roots.

More specifically, find how the conservation of energy (4.3) is transformed under the change of variables

$$x \to e_3 + \frac{(e_1 - e_3)(e_2 - e_3)}{y - e_3}.$$

Then, find where the segments $(-\infty, e_3]$, $[e_3, e_2]$, $[e_2, e_1]$ and $[e_1, +\infty)$ are mapped through this coordinate transformation.

Finally, show that the above imply that the period of the oscillatory motion and the "time of flight" of the scattering solution with the same energy are equal.

4.4 Find the energies for which the solution to the pendulum problem is expressed in terms of the Weierstrass elliptic function with a double root. Find the special expressions for the pendulum motion at these limits and verify that in one of the limits the solution degenerates to the stable equilibrium and simultaneously the period of motion equals the period of the small oscillations, $2\pi/\omega$.

4.5 For the motion of a particle in a hyperbolic potential, in the case $s = -1, t = -1$ and $E > -\omega^2$, which corresponds to the case of transmitting scattering solutions by a repulsive potential barrier, show that the motion is symmetric around the instant $t = 0$, namely show that $x(-t) = -x(t)$.

Obviously in the case of reflecting scattering solutions, $E < -\omega^2$, the symmetry of the problem implies that $x(-t) = x(t)$. What has to change in your previous derivation in this case?

4.6 For the motion of a particle in a hyperbolic potential, in the case $s = +1, t = -1$, which corresponds to oscillatory motion, the bounds of motion in Table 4.6 look asymmetric. However, the fact that the potential is even suggests that they should be symmetric. Verify that they actually are. Furthermore, verify that $x(t + T/2) = -x(t)$, where T is the period of the oscillation.

Show that at the double root limit of the solution, the solution degenerates to the equilibrium solution and that the period of motion tends to the period of the small oscillations $2\pi/\omega$.

References

1. H. Bateman, A. Erdélyi, *Higher Transcendental Functions*, vols. 1, 2, 3 (1953). ISBN 0-486-44614-X, ISBN 0-486-44615-8, ISBN 0-486-44616-6
2. M. Abramowitz, I. Stegun, *Handbook of Mathematical Functions with Formulas, Graphs, and Mathematical Tables* (United States Department of Commerce, National Bureau of Standards (NBS), 1964)

Chapter 5
Applications in Quantum Mechanics

Abstract In this chapter, we turn our attention to applications of the Weierstrass elliptic and related functions in Quantum Mechanics. It is well-known that Bloch's theorem implies that periodic potentials have in general a non-trivial band structure. However, it is not simple to find analytically solvable periodic potentials that demonstrate this behaviour. In this chapter, we focus on the n = 1 Lamé potential, which is a one-dimensional periodic potential that can be expressed in terms of the Weierstrass elliptic function. Its band structure can be analytically specified, using the properties of the Weierstrass functions.

There are many more interesting problems in classical mechanics that can be treated with the help of the Weierstrass elliptic function, than those that we studied in the previous chapter. We may refer to the point particle in a quartic potential, the spherical pendulum and the symmetric top as three such examples. I believe that the reader, having studied the cubic potential, the simple pendulum and the hyperbolic potentials, is prepared well to study any of these problems.

At his point we change the field of applications of the Weierstrass elliptic function, in order to demonstrate that these are not limited in the field of classical mechanics. We focus on one-dimensional periodic potentials in quantum mechanics. In particular we study a potential, which can be expressed in terms of the Weierstrass elliptic function, whose band structure can be analytically studied.

It is well known that particles that move in a periodic potential

$$V(x + a) = V(x) \tag{5.1}$$

accept as eigenfunctions, Bloch wave solutions of the form

$$\psi(x) = e^{ikx} u(x), \tag{5.2}$$

where $u(x)$ is periodic with the same period as the potential

$$u(x + a) = u(x). \tag{5.3}$$

G. Pastras, *The Weierstrass Elliptic Function and Applications in Classical and Quantum Mechanics*, SpringerBriefs in Physics,
https://doi.org/10.1007/978-3-030-59385-8_5

The parameter k is a function of the energy. When k is real, these wavefunctions are delta-function normalizable, while when k contains an imaginary part these wavefunctions are exponentially diverging. *This behaviour leads to the formation of a band structure*, i.e. the energy of the particles is not arbitrary but it is allowed to take values only within specific intervals. The reader may find more on Bloch's theorem at any sufficiently advanced quantum mechanics book, e.g. [3].

Although these are well known facts, it is not simple to find an analytically solvable periodic potential that demonstrates the formation of an non-trivial band structure.

For this chapter, it is crucial that the reader is familiar with the material of the first three chapters, especially the second one. These cover all the necessary background material. The handbooks [1, 2], which provide fast access to formulae and various properties may prove handy.

5.1   The $n = 1$ Lamé Potential

5.1.1   The $n = 1$ *Lamé Potential and Its Solutions*

Let's consider the periodic potential

$$V(x) = 2\wp(x),\tag{5.4}$$

where it is assumed that the moduli g_2 and g_3 are real. The Schrödinger equation reads

$$-\frac{d^2 y}{dx^2} + 2\wp(x)\, y = \lambda y.\tag{5.5}$$

This is the so called $n = 1$ Lamé equation—for general n, the potential is given by $V(x) = n(n+1)\wp(x)$. Historically, this equation was studied by Lamé towards completely different applications [4]. An excellent treatment of this class of ordinary differential equations in given by Ince in [5].

Consider the functions

$$y_\pm(x; a) = \frac{\sigma(x \pm a)}{\sigma(x)\, \sigma(\pm a)} e^{-\zeta(\pm a)x}.\tag{5.6}$$

It is easy to verify by direct computation that these functions are both eigenfunctions of the Schrödinger problem (5.5). Using the defining property of the Weierstrass ζ and σ functions (2.1) and (2.8), we find

$$\frac{dy_\pm}{dx} = (\zeta(x \pm a) - \zeta(x) \mp \zeta(\alpha))\, y_\pm,$$

$$\frac{d^2 y_\pm}{dx^2} = \left[(\zeta(x \pm a) - \zeta(x) \mp \zeta(\alpha))^2 - (\wp(x \pm a) - \wp(x))\right] y_\pm.$$

Then, applying the addition theorems of the Weierstrass $\wp$ and ζ functions (2.18) and (2.20), the above relations can be simplified to the form

$$\frac{dy_\pm}{dx} = \frac{1}{2} \frac{\wp'(x) \mp \wp'(a)}{\wp(x) - \wp(a)} y_\pm, \tag{5.7}$$

$$\frac{d^2 y_\pm}{dx^2} = (2\wp(x) + \wp(a)) y_\pm. \tag{5.8}$$

The last equation implies that $y_\pm$ are both eigenfunctions of the problem (5.5), corresponding to the eigenvalue

$$\lambda = -\wp(a). \tag{5.9}$$

As long as the eigenfunction modulus a is not equal to any of the three half-periods, the two σ functions appearing to the numerator of $y_\pm$ do not have roots at congruent positions. As such, the two wavefunctions are linearly independent and they provide the general solution. When the modulus a equals any of the half-periods though, it turns out that

$$y_\pm\left(x; \omega_{1,2,3}\right) = \sqrt{\wp(x) - e_{1,3,2}}. \tag{5.10}$$

For those eigenvalues, there is a second linearly independent solution, which is given by

$$\tilde{y}\left(x; \omega_{1,2,3}\right) = \sqrt{\wp(x) - e_{1,3,2}} \left(\zeta\left(x + \omega_{1,2,3}\right) + e_{1,3,2} x\right). \tag{5.11}$$

5.1.2 Reality of the Solutions

We would like to study whether the eigenfunctions (5.6) are real or not. Firstly, we consider the case of three real roots. In this case, $\wp(a)$ will assume all real values if a runs in the perimeter of the rectangle with corners located at 0, ω_1, ω_2 and ω_3. Since the Weierstrass elliptic function is of order two, for every point in the perimeter of this rectangle there is another point in the fundamental period parallelogram, where $\wp$ assumes the same value. Due to $\wp$ being an even function, this point is congruent to the opposite of the initial one. Therefore, the selection of the other point does not correspond to new eigenfunctions, but simply corresponds to the reflection $y_+ \leftrightarrow y_-$. Thus, it suffices to divide our analysis to four cases, one for each side of the rectangle with corners at the origin and the half-periods. In the following, b is considered always real.

1. a lies in the segment $[0, \omega_1]$. Then, $a = b$ and $\wp(a) > e_1$. In this case, trivially, the eigenfunctions (5.6) are real as,

$$\overline{y_\pm(x; b)} = y_\pm\left(x; \bar{b}\right) = y_\pm(x; b). \tag{5.12}$$

2. a lies in the segment $[0, \omega_2]$. Then, $a = ib$ and $\wp(a) < e_3$. In this case, trivially, the eigenfunctions (5.6) are complex conjugate to each other as,

$$\overline{y_\pm(x; ib)} = y_\pm\left(x; \overline{ib}\right) = y_\pm(x; -ib) = y_\mp(x; ib). \tag{5.13}$$

3. a lies in the segment $[\omega_2, \omega_3]$. Then, $a = \omega_2 + b$ and $e_3 < \wp(a) < e_2$. In this case,

$$\overline{y_\pm(x; \omega_2 + b)} = y_\pm\left(x; \overline{\omega_2 + b}\right) = y_\pm(x; -\omega_2 + b).$$

We use the quasi-periodicity properties of functions ζ (2.5) and σ (2.12) to find that

$$
\begin{aligned}
y_\pm(x; -\omega_2 + b) &= \frac{\sigma(x \mp \omega_2 \pm b)}{\sigma(x)\,\sigma(\mp\omega_2 \pm b)} e^{\mp\zeta(-\omega_2+b)x} \\
&= \frac{-\sigma(x \pm \omega_2 \pm b)\, e^{\mp 2\zeta(\omega_2)(x \pm \omega_2 \pm b \mp \omega_2)}}{-\sigma(x)\,\sigma(\pm\omega_2 \pm b)\, e^{\mp 2\zeta(\omega_2)(\pm\omega_2 \pm b \mp \omega_2)}} e^{\mp(\zeta(\omega_2+b)-2\zeta(\omega_2))x} \\
&= \frac{\sigma(x \pm \omega_2 \pm b)}{\sigma(x)\,\sigma(\pm\omega_2 \pm b)} e^{\mp\zeta(\omega_2+b)x} = y_\pm(x; \omega_2 + b),
\end{aligned}
$$

implying that

$$\overline{y_\pm(x; \omega_2 + b)} = y_\pm(x; \omega_2 + b). \tag{5.14}$$

Therefore, in this case the eigenfunctions (5.6) are real.

4. a lies in the segment $[\omega_1, \omega_3]$. Then, $a = \omega_1 + ib$ and $e_2 < \wp(a) < e_1$. In this case,

$$\overline{y_\pm(x; \omega_1 + ib)} = y_\pm\left(x; \overline{\omega_1 + ib}\right) = y_\pm(x; \omega_1 - ib).$$

As in previous case, we use the quasi-periodicity properties of ζ and σ to find

$$
\begin{aligned}
y_\pm(x; \omega_1 - ib) &= \frac{\sigma(x \pm \omega_1 \mp ib)}{\sigma(x)\,\sigma(\pm\omega_1 \mp ib)} e^{\mp\zeta(\omega_1 - ib)x} \\
&= \frac{-\sigma(x \mp \omega_1 \mp ib)\, e^{\pm 2\zeta(\omega_1)(x \mp \omega_1 \mp ib \pm \omega_1)}}{-\sigma(x)\,\sigma(\mp\omega_1 \mp ib)\, e^{\pm 2\zeta(\omega_1)(\mp\omega_1 \mp ib \pm \omega_1)}} e^{\mp(\zeta(-\omega_1 - ib)+2\zeta(\omega_1))x} \\
&= \frac{\sigma(x \mp \omega_1 \mp ib)}{\sigma(x)\,\sigma(\mp\omega_1 \mp ib)} e^{\pm\zeta(\omega_1 + ib)x} = y_\mp(x; \omega_1 + ib),
\end{aligned}
$$

meaning that

$$\overline{y_\pm(x; \omega_1 + ib)} = y_\mp(x; \omega_1 + ib). \tag{5.15}$$

In this case, the eigenfunctions (5.6) are complex conjugate to each other.

In the case of one real root, the situation is much simpler. $\wp(a)$ will assume all real values if a runs in the union of two segments, one on the real axis with endpoints 0 and $\omega_1 + \omega_2$ and one in the imaginary axis with endpoints 0 and $\omega_1 - \omega_2$. Similarly to the case of three real roots, there are more points where $\wp$ assumes real values, but their selection corresponds simply to the reflection $y_+ \leftrightarrow y_-$.

1. a lies in the segment $[0, \omega_1 + \omega_2]$. Then, $a = b$ and $\wp(a) > e_2$. This case is identical to the first case above, and, thus, the eigenfunctions (5.6) are real.
2. a lies in the segment $[0, \omega_1 - \omega_2]$. Then, $a = ib$ and $\wp(a) < e_2$. This case is identical to the second case above, and, thus, the eigenfunctions (5.6) are complex conjugate to each other.

5.1.3 The Band Structure of the $n = 1$ Lamé Potential: Three Real Roots

Comparing to the trivial case of a flat potential, we expect that when the eigenfunctions (5.6) are complex conjugate to each other, they are delta function normalizable Bloch waves, whereas when the eigenfunctions (5.6) are real, they are exponentially diverging non-normalizable states. However, in order to explicitly show that, we need to find how the eigenfunctions (5.6) transform under a shift of their argument by a period of the potential. For this purpose, we need to write the eigenfunctions (5.6) in the form (5.2). Let us consider the case of three real roots. Then, the period of the potential equals $2\omega_1$. Using the quasi-periodicity property of Weierstrass sigma function we get,

$$\frac{\sigma(x \pm a + 2\omega_1)}{\sigma(x + 2\omega_1)\sigma(\pm a)} = \frac{-e^{2\zeta(\omega_1)(x \pm a + \omega_1)}\sigma(x \pm a)}{-e^{2\zeta(\omega_1)(x+\omega_1)}\sigma(x)\sigma(\pm a)} = e^{\pm 2a\zeta(\omega_1)}\frac{\sigma(x \pm a)}{\sigma(x)\sigma(\pm a)}.$$

Thus, we may write the eigenfunctions (5.6) as

$$y_\pm(x; a) = u_\pm(x; a) e^{\pm ik(a)x}, \tag{5.16}$$

where

$$u_\pm(x; a) = \frac{\sigma(x \pm a)}{\sigma(x)\sigma(a)} e^{\mp \frac{a\zeta(\omega_1)}{\omega_1}x}, \tag{5.17}$$

$$ik(a) = \frac{a\zeta(\omega_1) - \omega_1\zeta(a)}{\omega_1} \tag{5.18}$$

and it holds that $u_\pm(x + 2\omega_1; a) = u_\pm(x; a)$.

In order to assort these states into non-normalizable states and Bloch waves, we need to study the function

$$f(a) = a\zeta(\omega_1) - \omega_1\zeta(a). \tag{5.19}$$

This is clearly not an elliptic function, but rather a quasi-periodic function, since it obeys,

$$f(a + 2\omega_1) = (a + 2\omega_1)\zeta(\omega_1) - \omega_1(\zeta(a) + 2\zeta(\omega_1)) = f(a), \tag{5.20}$$
$$f(a + 2\omega_2) = (a + 2\omega_2)\zeta(\omega_1) - \omega_1(\zeta(a) + 2\zeta(\omega_2)) = f(a) + i\pi. \tag{5.21}$$

However, its derivative equals

$$f'(a) = \zeta(\omega_1) + \omega_1 \wp(a), \tag{5.22}$$

which is clearly an order two elliptic function. Therefore, the function $f(a)$ is stationary exactly twice in each cell. Furthermore, $f'(a)$ is real, wherever $\wp(a)$ is real, thus, everywhere in the space where the modulus a takes values.

Recalling the property (2.7), we can show that the function $f(a)$ takes the following values at the origin and the half-periods:

$$\lim_{b \to 0^+} f(b) = - \lim_{b \to 0^+} \frac{\omega_1}{b} = -\infty, \tag{5.23}$$

$$\lim_{b \to 0^+} f(ib) = - \lim_{b \to 0^+} \frac{\omega_1}{ib} = +i\infty, \tag{5.24}$$

$$f(\omega_1) = \omega_1 \zeta(\omega_1) - \omega_1 \zeta(\omega_1) = 0, \tag{5.25}$$

$$f(\omega_2) = \omega_2 \zeta(\omega_1) - \omega_1 \zeta(\omega_2) = i\frac{\pi}{2}, \tag{5.26}$$

$$f(\omega_3) = (\omega_1 + \omega_2)\zeta(\omega_1) - \omega_1 \zeta(\omega_1 + \omega_2) = i\frac{\pi}{2}. \tag{5.27}$$

Since the derivative of $f(a)$ is real at the perimeter of the rectangle with corners the origin and the three half-periods, it follows that on the sides $[0, \omega_1]$ and $[\omega_2, \omega_3]$ only the real part of $f(a)$ varies, whereas on the sides $[0, \omega_2]$ and $[\omega_1, \omega_3]$ only the imaginary part of $f(a)$ varies.

Since the real part of $f(a)$ is identical at $a = \omega_2$ and $a = \omega_3$ (it vanishes), the mean value theorem implies that there is a point in the segment $[\omega_2, \omega_3]$ where the derivative of $f(a)$ vanishes. In every cell, there is another point, where the derivative vanishes, which is congruent to the opposite of the above point, and, thus, it is not congruent to any point of the perimeter of the rectangle with corners at the origin and the half-periods. Since the derivative of $f(a)$ is an order two elliptic function, there is no other point in a cell, and, thus, in the aforementioned rectangle, where $f(a)$ is stationary.

The above, combined with the values of $f(a)$ at the origin and the half-periods, imply that:

1. At the segment $[0, \omega_1]$, $f(a)$ is everywhere real. It is nowhere stationary in this segment and therefore, it varies monotonically from $-\infty$ at the origin to 0 at ω_1. It follows that it vanishes nowhere except at $a = \omega_1$.
2. At the segment $[\omega_1, \omega_3]$, $f(a)$ is everywhere purely imaginary. It is nowhere stationary in this segment and therefore, its imaginary part varies monotonically from 0 at ω_1 to $\pi/2$ at ω_3.
3. At the segment $[\omega_2, \omega_3]$, $f(a)$ has an imaginary part equal to $\pi/2$. The real part vanishes at the endpoints of the segment, it reaches a minimum value at the stationary point of $f(a)$ and it vanishes nowhere expect at the endpoints, since the derivative of $f(a)$ vanishes only once.

4. At the segment $[0, \omega_2]$, $f(a)$ is everywhere purely imaginary. It is nowhere stationary in this segment and therefore, its imaginary part varies monotonically from $+\infty$ at the origin to $\pi/2$ at ω_2.

It follows that $k(a)$ is purely imaginary, as required for Bloch waves, in the segments $[0, \omega_2]$ and $[\omega_1, \omega_3]$ and nowhere else. Thus, *the band structure of the $n = 1$ Lamé potential, in the case of three real roots, contains a finite "valence" band between the energies $-e_1$ and $-e_2$ and an infinite "conduction" band for energies larger than $-e_3$.* The former corresponds to wavefunctions with a parameter a taking values in the segment $[\omega_1, \omega_3]$, whereas the latter corresponds to wavefunctions with a parameter a taking values in the segment $[0, \omega_2]$.

5.1.4 The Band Structure of the $n = 1$ Lamé Potential: One Real Root

In the case of one real root, the period of the potential equals $2\omega_1 + 2\omega_2$. We write the eigenfunctions (5.6) as

$$y_\pm(x; a) = u_\pm(x; a) e^{\pm ik(a)x}, \tag{5.28}$$

where

$$u_\pm(x; a) = \frac{\sigma(x \pm a)}{\sigma(x)\sigma(a)} e^{\mp \frac{a\zeta(\omega_1+\omega_2)}{\omega_1+\omega_2}x}, \tag{5.29}$$

$$ik(a) = \frac{a\zeta(\omega_1+\omega_2) - (\omega_1+\omega_2)\zeta(a)}{\omega_1+\omega_2} \tag{5.30}$$

and $u_\pm(x + 2\omega_1 + 2\omega_2; a) = u_\pm(x; a)$.
We now have to study the function

$$f(a) = a\zeta(\omega_1+\omega_2) - (\omega_1+\omega_2)\zeta(a). \tag{5.31}$$

Similarly to the case of three real roots, the function $f(a)$ is quasi-periodic

$$f(a + 2\omega_1) = (a + 2\omega_1)\zeta(\omega_1+\omega_2) - (\omega_1+\omega_2)(\zeta(a) + 2\zeta(\omega_1))$$
$$= f(a) - i\pi, \tag{5.32}$$
$$f(a + 2\omega_2) = (a + 2\omega_2)\zeta(\omega_1+\omega_2) - (\omega_1+\omega_2)(\zeta(a) + 2\zeta(\omega_2))$$
$$= f(a) + i\pi, \tag{5.33}$$

whereas its derivative equals

$$f'(a) = \zeta(\omega_1+\omega_2) + (\omega_1+\omega_2)\wp(a), \tag{5.34}$$

which is an order two elliptic function. Simple arguments, similar to those we used in the case of three real roots, imply that $f(a)$ is stationary exactly once in the union of the segments $[0, \omega_1 + \omega_2]$ and $[0, \omega_1 - \omega_2]$. Furthermore, the derivative $f'(a)$ is real wherever $\wp(a)$ is real and therefore only the real part of $f(a)$ varies on the segment $[0, \omega_1 + \omega_2]$, whereas only the imaginary part of $f(a)$ varies on the segment $[0, \omega_1 - \omega_2]$. We can easily check that

$$\lim_{b \to 0^+} f(b) = -\lim_{b \to 0^+} \frac{\omega_1 + \omega_2}{b} = -\infty, \tag{5.35}$$

$$\lim_{b \to 0^+} f(ib) = -\lim_{b \to 0^+} \frac{\omega_1 + \omega_2}{ib} = +i\infty, \tag{5.36}$$

$$f(\omega_1 + \omega_2) = (\omega_1 + \omega_2)\zeta(\omega_1 + \omega_2) - (\omega_1 + \omega_2)\zeta(\omega_1 + \omega_2) = 0, \tag{5.37}$$

$$f(\omega_1 - \omega_2) = (\omega_1 - \omega_2)\zeta(\omega_1 + \omega_2) - (\omega_1 + \omega_2)\zeta(\omega_1 - \omega_2) = -i\pi, \tag{5.38}$$

and, thus, $f(a)$ is real on $[0, \omega_1 + \omega_2]$ and purely imaginary on $[0, \omega_1 - \omega_2]$.

The only remaining question to be answered is whether the stationary point of $f(a)$ belongs in $[0, \omega_1 + \omega_2]$ or $[0, \omega_1 - \omega_2]$. In the first case, there is a single point on $[0, \omega_1 + \omega_2]$, where $f(a)$ vanishes, and, thus, it can be considered purely imaginary, while in the second case there is not. It turns out that the stationary point lies in $[0, \omega_1 - \omega_2]$ and therefore $f(a)$ is purely imaginary everywhere on $[0, \omega_1 - \omega_2]$ and nowhere else. Thus, *the band structure of the $n = 1$ Lamé potential, in the case of one real root, contains only an infinite "conduction" band for energies larger than $-e_2$.*

5.2 The Bounded $n = 1$ Lamé Potential

5.2.1 The Bounded $n = 1$ Lamé Potential and Its Band Structure

If three real roots exist, the potential $V = 2\wp(x + \omega_2)$ is also real. Although the expressions for this potential and the one of the previous section, $V = 2\wp(x)$, look similar, physically they are very different. While the potential $V = 2\wp(x)$ is unbounded and it is singular at all points $x = 2n\omega_1$, $n \in \mathbb{Z}$, the potential $V = 2\wp(x + \omega_2)$ is bounded and smooth everywhere. We may repeat the work done in the previous section in order to determine the eigenstates and the band structure of the bounded potential.

Following the previous section, it is trivial to show that the functions

$$\psi_{\pm}(x; a) = y_{\pm}(x + \omega_2; a) = \frac{\sigma(x + \omega_2 \pm a)}{\sigma(x + \omega_2)\sigma(a)} e^{\mp\zeta(a)(x+\omega_2)},$$

which emerge directly from the eigenfunctions of the unbounded problem (5.6) via a shift of their argument by ω_2, obey

$$\frac{d^2\psi_\pm(x;a)}{dx^2} = (2\wp(x+\omega_2) + \wp(a))\,\psi_\pm(x;a)\,, \tag{5.39}$$

and, thus, they are both eigenfunctions of the bounded $n = 1$ Lamé problem, with an eigenfunction equal to

$$\lambda = \wp(a)\,. \tag{5.40}$$

However, this asymmetric insertion of ω_2 has deprived the eigenfunctions from their nice reality properties. For example, for real a,

$$\begin{aligned}
\bar\psi_+(x;a) &= \frac{\sigma(x-\omega_2+a)}{\sigma(x-\omega_2)\,\sigma(a)}e^{-\zeta(a)(x-\omega_2)}\\
&= \frac{-\sigma(x+\omega_2+a)\,e^{-2\zeta(\omega_2)(x+\omega_2+a-\omega_2)}}{-\sigma(x+\omega_2)\,\sigma(a)\,e^{-2\zeta(\omega_2)(x+\omega_2-\omega_2)}}e^{-\zeta(a)(x+\omega_2)}e^{2\zeta(a)\omega_2}\\
&= e^{2(\omega_2\zeta(a)-a\zeta(\omega_2))}\psi_+(x;a)\,,
\end{aligned}$$

implying that $\psi_+(x;a)$ is not real, but it has to be rotated by a constant complex phase in order to become so. It is left as an exercise to the reader to repeat the algebra of Sect. 5.1 and show that the eigenfunctions

$$\psi_\pm(x;a) = \frac{\sigma(x+\omega_2\pm a)\,\sigma(\omega_2)}{\sigma(x+\omega_2)\,\sigma(\omega_2\pm a)}e^{-\zeta(\pm a)x}\,, \tag{5.41}$$

have reality properties similar to those of the eigenfunctions of the unbounded Lamé potential (5.6). It is also left to the reader to verify that *the band structure of the bounded potential is identical to the band structure of the unbounded one.*

It is quite interesting that the potentials $V = 2\wp(x)$ and $V = 2\wp(x+\omega_2)$ have the same band structure. As we already commented, the two potentials are quite dissimilar functions, the unbounded one having second order poles, whereas the bounded one being smooth, as shown in Fig. 5.1.

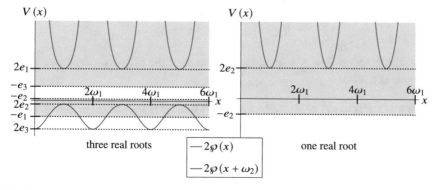

Fig. 5.1 The band structure of the $n = 1$ Lamé potential. The shadowed regions correspond to the allowed bands

5.2.2 Connection Between the Bounded and Unbounded $n = 1$ Lamé Problems

The fact that the unbounded and bounded Lamé potentials are isospectral is not a coincidence. Assume the function

$$W(x) = \frac{\wp'(x)}{2(\wp(x) - e_3)},$$

(5.42)

which we will call the "superpotential" and the operators

$$A = \frac{d}{dx} + W(x), \quad A^\dagger = -\frac{d}{dx} + W(x),$$

(5.43)

which we will call "annihilation" and "creation" operators respectively. Then,

$$A^\dagger A = -\frac{d^2}{dx^2} + V_1(x), \quad V_1(x) = W^2(x) - W'(x),$$

(5.44)

$$AA^\dagger = -\frac{d^2}{dx^2} + V_2(x), \quad V_2(x) = W^2(x) + W'(x).$$

(5.45)

It is a matter of algebra to show that for the specific superpotential given by (5.42), it turns out that

$$V_1(x) = 2\wp(x + \omega_2) + e_3, \quad V_2(x) = 2\wp(x) + e_3.$$

(5.46)

In other words,

$$A^\dagger A = \tilde{H} + e_3, \quad AA^\dagger = H + e_3,$$

(5.47)

where H is the Hamiltonian of the unbounded $n = 1$ Lamé problem and $\tilde{H}$ is the Hamiltonian of the bounded $n = 1$ Lamé problem.

Now consider an eigenfunction y of the unbounded problem with energy E, i.e.

$$Hy = \left(AA^\dagger - e_3\right)y = Ey.$$

(5.48)

Then, the function $\tilde{y} = A^\dagger y$ is an eigenfunction of the bounded problem with the same energy, since

$$\tilde{H}\tilde{y} = \left(A^\dagger A - e_3\right)A^\dagger y = A^\dagger\left(AA^\dagger - e_3\right)y = A^\dagger Hy = A^\dagger Ey = E\tilde{y}.$$

(5.49)

The above are general characteristics of partner potentials that emerge from a superpotential. If the reader finds this example attractive enough, so that they desire to learn more on supersymmetric quantum mechanics, they may consult [6, 7].

Action with the operator $A^\dagger$ cannot transform a Bloch wave to an exponentially diverging function or the other way around. As a result, the two Hamiltonians are isospectral. It is left as an exercise to the reader to show that indeed

$$A^\dagger y_\pm (x; a) = c \psi_\pm (x; a),\qquad(5.50)$$

where $y_\pm$ are given by Eq. (5.6), $\psi_\pm$ are given by Eq. (5.41) and c is a constant.

Problems

5.1 Show that the special solution (5.11) is indeed an eigenfunction of the $n = 1$ Lamé problem, corresponding to an eigenvalue equal to one of the three roots of the respective Weierstrass elliptic function.

5.2 In order to study the band structure of the $n = 1$ Lamé problem, we expressed the problem eigenfunctions as Bloch waves in the form of Eq. (5.16), where the functions u and k are given by Eqs. (5.17) and (5.18). If this is carried out properly, the function u will be periodic function with period equal to that of the potential. Verify that this is indeed correct.

5.3 Show that the eigenfunctions (5.41) of the bounded $n = 1$ Lamé potential have the same reality properties as the eigenfunctions (5.6) of the unbounded one.

5.4 Find how the eigenfunctions (5.41) of the bounded $n = 1$ Lamé potential transform under a shift of their argument by the period of the potential. Once you have accomplished that, deduce the band structure of the bounded $n = 1$ Lamé potential.

5.5 Using the addition theorem of the Weierstrass σ function (2.19), show that the eigenstates $y_\pm$, as given by Eq. (5.6) obey the following "'normalization" properties:

$$y_+ y_- = \wp (x) - \wp (a),\qquad(5.51)$$
$$y_+' y_- - y_+ y_-' = -\wp' (a).\qquad(5.52)$$

How are these properties modified for the eigenfunctions of the bounded $n = 1$ Lamé potential?

5.6 We have shown that the eigenfunctions of the $n = 1$ Lamé potential as given by Eq. (5.6), which correspond to the allowed bands are complex conjugate to each other. Therefore they can be written in terms of a magnitude and an argument like

$$y_\pm (x; a) = R (x; a) e^{\pm i \Phi(x;a)}. \tag{5.53}$$

Specify the square of the magnitude R^2 and the derivative of the argument $d\Phi/dx$ and verify that they are elliptic functions. You may find the results of problem 5.5 useful.

The fact that $d\Phi/dx$ is an elliptic function implies that the argument Φ is a quasi-periodic function with similar behaviour to that of the function ζ. Specify its quasi-periodicity property in the real axis. Compare with Eq. (5.18). Are the two expressions compatible?

5.7 Show that the action of the creation operator $A^\dagger$ which is given by Eq. (5.43), on the eigenstates of the unbounded $n = 1$ Lamé problem (5.6), results to the eigenstates of the bounded $n = 1$ Lamé problem (5.41). Namely, verify that Eq. (5.50) holds and specify the coefficient c. How does this coefficient compares with the results of problem 5.5?

5.8 At the limit that the two larger roots coincide, the real period diverges. Therefore, the $n = 1$ Lamé potential ceases being a periodic potential. Study the bounded potential at this limit. Find the special form of the wavefunctions and their corresponding eigenvalues. Then, focus on two special properties:

1. What is the form of the states of the finite allowed band at this limit? What is their normalization?
2. What is the form of the states of the infinite allowed band? What are the reflection and transmission coefficients?

The answers you got in the above questions are not a coincidence. Find the partner potentials that emerge from the superpotential $W = \sqrt{3e_0} \tanh \sqrt{3e_0}x$, where e_0 is the double root, as we did in Sect. 5.2.2, and compare with the double root limit of the bounded $n = 1$ Lamé potential. Can you deduce the above properties from this fact?

References

1. H. Bateman, A. Erdélyi, *Higher Transcendental Functions*, vols. 1, 2, 3 (1953). ISBN 0-486-44614-X, ISBN 0-486-44615-8, ISBN 0-486-44616-6
2. M. Abramowitz, I. Stegun, *Handbook of Mathematical Functions with Formulas, Graphs, and Mathematical Tables* (United States Department of Commerce, National Bureau of Standards (NBS), 1964)
3. L.E. Ballentine, *Quantum Mechanics: A Modern Development*, 2nd edn. (World Scientific Publishing Co., 2014). ISBN 978-9814578585

4. G. Lamé, *Sur les Surfaces Isothermes dans les Corps Homogènes en Équilibre de Température*. Journal de mathématiques pures et appliquées **2**, 147–188. Available at Gallica (1837)
5. E.L. Ince, *Ordinary Differential Equations* (Dover Publications, New York, 1944). ISBN 978-0-486-60349-0
6. F. Cooper, A. Khare, U. Sukhatme, *Supersymmetry and Quantum Mechanics*, Physics Reports 251, pp. 267–385 (1995)
7. A. Gangopadhyaya, J.V. Mallow, C. Rasinariu, *Supersymmetric Quantum Mechanics: An Introduction* (World Scientific Publishing Co., 2010). ISBN 978-9814313094

Chapter 6
Epilogue and Projects for the Advanced Reader

Abstract Chapter 5 concludes our short but thorough introduction to the Weierstrass elliptic function and its applications in Physics. In this last chapter, we provide three more exercises-projects for the curious and advanced reader. They are inspired from contemporary research in integrable systems, in particular the sine-Gordon equation, as well as string theory, in particular the study of classical string solutions on symmetric spaces.

The presentation of the $n = 1$ Lamé problem concludes this short introduction to the Weierstrass elliptic function and its applications in Physics. The applications of elliptic functions are not limited to this content.

Many more problems in classical mechanics can be analytically solved in terms of elliptic functions. We may refer to the point particle in a quartic potential, the spherical pendulum and the symmetric top as three such examples. Furthermore, in quantum mechanics, potentials of the form

$$V(x) = n(n+1) \wp(x), \tag{6.1}$$

where $n \in \mathbb{Z}$ can be analytically solved in terms of elliptic functions and present amazing features. Such potentials have a richer band structure containing up to n finite gaps in their spectrum.

The Lamé equation also appears in other problems of classical Physics, whenever one expresses the Laplace operator in elliptical coordinates. Actually, this is the historical reason for the study of this class of linear differential equations. For example, elliptic functions will appear if one studies the heat diffusion on the surface of an ellipsoid by rotation, such as the earth surface.

Furthermore, elliptic functions find applications in many fields of more modern physics. Several solutions of very interesting integrable systems, such as the sine-Gordon, the sinh-Gordon or the Korteweg-de Vries (KdV) equation can be analytically expressed in terms of elliptic functions. Such solutions can be further used to

G. Pastras, *The Weierstrass Elliptic Function and Applications in Classical and Quantum Mechanics*, SpringerBriefs in Physics,
https://doi.org/10.1007/978-3-030-59385-8_6

construct analytic string solutions in symmetric spaces, such as dS and AdS spaces, as well as minimal surfaces in hyperbolic spaces. Elliptic functions also emerge naturally when one calculates string scattering amplitudes for world-sheets that have the topology of a torus.

Extending this book to these directions is beyond its scope, which is to provide the students a fast but thorough introduction to the subject, without the need to attend a whole mathematics course. However, for the curious and advanced reader, we add three more exercises-projects, which are inspired from current-date research in integrable systems and string theory.

6.1 Building Sine-Gordon Kinks on Top of an Elliptic Background

The sine-Gordon equation is a relativistic two-dimensional system, which exhibits an incredibly rich structure and finds applications in a large variety of sectors in physics. The sine-Gordon equation reads

$$-\frac{\partial^2 \varphi}{\partial t^2} + \frac{\partial^2 \varphi}{\partial x^2} \equiv \frac{\partial^2}{\partial x_+ \partial x_-}\varphi = \mu^2 \sin \varphi. \tag{6.2}$$

The coordinates $x_\pm$ are defined so that $x_\pm = \frac{1}{2}(x \pm t)$. This equation can be conceived as the continuous limit of a sequence of coupled pendulums.

This equation accepts an infinite set of vacuum solutions of the form

$$\varphi = 2n\pi, \quad n \in \mathbb{Z}. \tag{6.3}$$

If you are familiar with elementary field theory or Noether's theorem, it is not difficult to show that the energy and momentum densities are given by

$$\mathcal{E} = \frac{1}{2}\left(\frac{\partial \varphi}{\partial t}\right)^2 + \frac{1}{2}\left(\frac{\partial \varphi}{\partial x}\right)^2 - \mu^2 \cos \varphi, \tag{6.4}$$

$$\mathcal{P} = -\frac{\partial \varphi}{\partial t}\frac{\partial \varphi}{\partial x}. \tag{6.5}$$

One fundamental property of the sine-Gordon equation is the fact that its solutions are connected in pairs, through the first-order differential equations,

$$\frac{\partial}{\partial x_+}\frac{\varphi + \tilde{\varphi}}{2} = a\mu \sin \frac{\varphi - \tilde{\varphi}}{2}, \tag{6.6}$$

$$\frac{\partial}{\partial x_-}\frac{\varphi - \tilde{\varphi}}{2} = \frac{1}{a}\mu \sin \frac{\varphi + \tilde{\varphi}}{2}, \tag{6.7}$$

which are called Bäcklund transformations.

These transformations can be used in order to build solutions out of a given one, the so called seed solution, without the need of solving second-order equations, but rather solving first-order ones. The archetypical example is the use of the vacuum solution $\varphi = 0$ as the seed solution. This leads to the construction of the solutions

$$\varphi = 4 \arctan e^{\mu\left(ax_+ + \frac{x_-}{a}\right)+c} = 4 \arctan e^{\mu\gamma(x-vt)+c}, \tag{6.8}$$

where $v = \frac{1-a^2}{1+a^2}$ and $\gamma = 1/\sqrt{1-v^2}$. These are the so called kink solutions. These solutions are of topological nature. They look like a twist of the field that connects two adjacent vacua.

Using the kink solutions as the seed for a subsequent Bäcklund transformation leads to two-kink solutions and so on. Such solutions describe multiple kinks interacting with each other. It turns out that kinks are never reflected, however the encounter of two kinks causes a specific time-delay to their propagation.

For a more detailed introduction to the sine-Gordon equation, the reader is referred to [1, 2].

One may consider solutions of the sine-Gordon equation that depend solely on time or position. In a trivial manner, in such cases, the sine-Gordon equation (6.2) degenerates to the equation of the simple pendulum. It follows that such solutions can be expressed in terms of the Weierstrass elliptic function, as in Sect. 4.2.

1. Write down the solutions of the sine-Gordon equation that depend solely on time or position in terms of the Weierstrass elliptic function. Study the energy and momentum densities for these solutions.
2. The apparent advantage of using the elliptic solutions as seeds in the Bäcklund transformations is the fact that they depend on either x or t solely. Express the Bäcklund transformations (6.6) and (6.7) in terms of the time t and position x instead of the coordinates $x_\pm$.
3. For the moment stick to the solutions that depend solely on time. Calculate the squares of the quantities $\cos\frac{\varphi}{2}$, $\sin\frac{\varphi}{2}$ and $\frac{\partial\varphi}{\partial t}$ that appear in the Bäcklund transformations in the form that you specified in step 2. Their signs are not going to play a crucial role, but you can specify them; in general they behave differently in the case of oscillating and rotating solutions.
4. Use the elliptic solutions as seeds in the Bäcklund transformations. You must notice that one of the two Bäcklund transformations has turned into an equation that contains only the derivative with respect to the position and all other functions that appear are functions of time solely. Therefore, this can be solved as an ordinary differential equation, where the undetermined integration constants are unknown functions of time. For simplicity assume that the parameter a is positive and larger than one.
5. Once you have solved the above equation, you may substitute the solution to the remaining Bäcklund transformation to find the new solutions. In the process you may find useful the definitions of two constants, namely $D^2 = \frac{1}{4}\left[\mu^2\left(a^2+a^{-2}\right)-2E\right]$ and $\tilde{a}$, which is defined so that $\wp(\tilde{a}) = -\frac{E}{6} + \frac{\mu^2}{4}\left(a^2+a^{-2}\right)$. You may also need to use the results of problem 5.6.

Check what is the sign of the parameter D^2. Does it depend on whether the seed was representing an oscillating or a rotating solution of the pendulum?

6. Stick to the case $D^2 > 0$. The new solutions that you have constructed describe kinks, i.e. twists of the field that connect adjacent vacua, propagating on top of an excited background. This is clearer when you consider seeds that correspond to oscillating solutions of the pendulum. The position of the kink is the center of the area where the field performs the jump from the region of the one vacuum to the region of the adjacent one. Locate the position of the kink and study its velocity. In general this velocity is varying with time, but you will be able to calculate a well-defined mean velocity. Draw some plots of the velocity as a function of the parameter $\tilde{a}$ in order to specify whether it is subluminal or superluminal.

7. Find how the solution looks like asymptotically far away from the kink. You should find that the solution is similar to the seed. What is the effect on this position-independent background oscillation that the elliptic solution is, by the passage of the kink? You just found the physical importance of the parameter $\tilde{a}$ that you defined above.

8. Find the energy and momentum of the kink. At first you may think that the former is infinite, but this is due to the fact that the seed solution has also infinite energy. Find the energy of the kink integrating the difference of the energy densities of the new and seed solutions. You just found the physical importance of the parameter D that you defined above.

9. Find how the energy of the kink is connected to the effect that it causes on the background motion of the system. This is an "equation of state" of the system that in principle can be verified experimentally.

10. You may repeat/modify steps 3 to 9 for the sine-Gordon solutions that depend solely on position.

If you meet an obstacle that you cannot overcome at any of the steps of the problem, you may look for hints in [3, 4].

6.2 Building Classical String Solutions on the Sphere

Consider a classical relativistic string propagating in flat spacetime. As it moves, it carves a two-dimensional sheet, which is usually called the world-sheet of the string. We parametrize this world-sheet using two parameters; one space-like parameter x^1 and one time-like parameter x^0. You may imagine that the space-like parameter identifies a point of the string, whereas the time-like parameter is the physical time, but this is not necessarily exactly true. In any case, this parametrization provides an embedding of the string world-sheet into the flat spacetime via the embedding functions $X^\mu = X^\mu \left(x^0, x^1 \right)$. The dynamics of the relativistic string are determined by demanding that the area of the world-sheet is stationary, i.e. the action of the system is

$$S = T \int dx^0 dx^1 \sqrt{-\det \gamma_{ij}}, \tag{6.9}$$

where $i, j = 0, 1$, the constant T is the tension of the string and the 2×2 matrix $\gamma_{ij} = \sum_{\mu,\nu} \eta_{\mu\nu} \frac{\partial X^\mu}{\partial x^i} \frac{\partial X^\nu}{\partial x^j}$ is the induced metric on the world-sheet. The matrix $\eta_{\mu\nu}$ is equal to $\eta_{\mu\nu} = \text{diag}\{-1, +1, \ldots, +1\}$ and it is the metric of the Minkowski space. This is called the Nambu-Goto action.

One can introduce an auxiliary metric field g, which allows the re-expression of the action in a more handy form, namely,

$$S = T \int dx^0 dx^1 \sqrt{-\det g_{ij}} \sum_{a,b,\mu,\nu} \eta_{\mu\nu} g^{ab} \frac{\partial X^\mu}{\partial x^a} \frac{\partial X^\nu}{\partial x^b}, \tag{6.10}$$

where g^{ab} is the inverse matrix of g_{ab}. This is the so-called Polyakov action. It is completely equivalent to the Nambu-Goto action; one can show this by substituting the auxiliary field g_{ab} using its equations of motion. Both actions (6.9) and (6.10) have a huge reparametrization symmetry. They are invariant under diffeomorphisms, i.e. the redefinition of the world-sheet coordinates x^0 and x^1, $x^0 \to x'^0 (x^0, x^1)$, $x^1 \to x'^1 (x^0, x^1)$.

Apart from the equations of motion for the embedding functions X^μ, when one uses the Polyakov action, they have to take into consideration the equations of motion for the auxiliary metric field. These are the so-called Virasoro constraints.

One may use part of the diffeomorphism invariance to define world-sheet coordinates $x^\pm$, so that the diagonal elements of the fiducial metric g vanish. Then, the Polyakov action assumes the form

$$S = T \int dx^+ dx^- \sum_{\mu,\nu} \eta_{\mu\nu} \frac{\partial X^\mu}{\partial x^+} \frac{\partial X^\nu}{\partial x^-}. \tag{6.11}$$

Such a selection does not completely fix the world-sheet coordinates. The action is still invariant under redefinitions of the form $x^+ \to x'^+ (x^+)$, $x^- \to x'^- (x^-)$. You may find more details on string actions in [5].

One can describe a string moving in a submanifold of the original flat spacetime by introducing a Lagrange multiplier term with the appropriate geometric constraint. For example a string moving on the surface of a two-dimensional sphere is described by the action

$$S = T \int dx^+ dx^- \left[\sum_{\mu,\nu} \eta_{\mu\nu} \frac{\partial X^\mu}{\partial x^+} \frac{\partial X^\nu}{\partial x^-} - \lambda \left(\sum_{j=1}^{3} (X^j)^2 - R^2 \right) \right]. \tag{6.12}$$

The equation of motion for X^0 assumes the form

$$\frac{\partial^0 X^0}{\partial x^+ \partial x^-} = 0 \Rightarrow X^0 = f_+(x_+) + f_-(x_-). \tag{6.13}$$

We may take advantage of the string world-sheet re-parametrization symmetry to re-define the coordinates $x^\pm$, so that

$$f_\pm(x^\pm) = m_\pm x'^\pm \Rightarrow X^0 = m_+ x^+ + m_- x^-. \tag{6.14}$$

The equations of motion for X^i, $i = 1, 2, 3$ assume the form

$$\frac{\partial^0 X^i}{\partial x^+ \partial x^-} = -\frac{1}{R^2} \left(\sum_{j=1}^{3} \frac{\partial X^j}{\partial x^+} \frac{\partial X^j}{\partial x^-} \right) X^i, \tag{6.15}$$

while the equation for the Lagrange multiplier is written as

$$\sum_{j=1}^{3} \left(X^j \right)^2 = R^2. \tag{6.16}$$

Finally, one has also to satisfy the Virasoro constraints

$$\sum_{j=1}^{3} \frac{\partial X^j}{\partial x^+} \frac{\partial X^j}{\partial x^+} = m_+^2, \quad \sum_{j=1}^{3} \frac{\partial X^j}{\partial x^-} \frac{\partial X^j}{\partial x^-} = m_-^2. \tag{6.17}$$

When the string is propagating in a symmetric space, like the sphere, the equations above have a very interesting property. There is a non-trivial, non-local coordinate transformation which transforms them to the equations of an integrable system of the family of the sine-Gordon equation that you met in the previous problem. This is the so-called Pohlmeyer reduction. You may find the original work in [6]. In the case of a string propagating on a two-dimensional sphere, this is the sine-Gordon equation itself. Indeed if we define the Pohlmeyer field φ as,

$$\sum_{j=1}^{3} \frac{\partial X^j}{\partial x^+} \frac{\partial X^j}{\partial x^-} := m_+ m_- \cos \varphi, \tag{6.18}$$

you may show that the equations of motion imply that φ obeys

$$\frac{\partial^2 \varphi}{\partial x^+ \partial x^-} = \mu^2 \sin \varphi, \tag{6.19}$$

where $\mu^2 = -m_+ m_- / R^2$

Although it is trivial to find the solution of the sine-Gordon equation that corresponds to a given string solution, the converse is highly non-trivial, due to the nonlocality of the Pohlmeyer field definition. However, if you select solutions of the sine-Gordon equation that depend only on either the space-like parameter $x^1 = x^+ + x^-$ or the time-like parameter $x^0 = x^+ - x^-$ this task is possible.

1. Go back to the previous problem, and write down the solutions of the sine-Gordon equation that depend solely on time or position. As you already know, these are connected to the solution of the simple pendulum and they are expressed in terms of the Weierstrass elliptic function. There are four classes of them: solutions that depend only on time and solutions that depend only on space; for each of these two classes, there are solutions that correspond to oscillating motions of the pendulum and solutions that correspond to rotating motions of the pendulum.

2. Use the definition of the Pohlmeyer field, in order to write down the equations of motion for X^1, X^2 and X^3, as *linear* differential equations with varying coefficients, which are expressed in terms of the Pohlmeyer field. Substitute the sine-Gordon solutions from the previous step. Notice that these differential equations can be solved via separation of variables. Separate the variables and result in two effective Schrödinger problems with connected eigenvalues. You will meet a well-known friend from Chap. 5, the $n = 1$ Lamé equation. Categorize the solutions into three classes depending on whether the eigenvalue of the trivial problem is positive, negative or vanishing. Bear in mind that the eigenfunctions of both effective Schrödinger problems do not have the interpretation of a wavefunction, and thus, they can be non-normalizable states in general.

3. By now you know the general solution to the equations of motion. Each embedding function X^j is a linear combination of the functions you specified above for all possible eigenvalues. However it is left to satisfy the geometric constraint (6.16), as well as the Virasoro constraints (6.17). The form of the geometric constraint strongly suggests that you should form an ansatz, where the first two components are the general solution corresponding to a single positive eigenvalue of the trivial effective problem, whereas the third component is given by the special case of a vanishing eigenvalue of the latter. Demand that the solution is real, what does this imply for the eigenstates of the $n = 1$ Lamé problem? Should they be Bloch waves or non-normalizable states? Find what the geometric constraint implies for the solution. You will find the results of the problem 5.5 useful.

4. Finally, express the Virasoro constraints (6.17) in terms of the world-sheet coordinates x^0 and x^1. Use the geometric constraint and by parts integration to write them so that they contain the second derivatives of the embedding functions. The latter can be deduced directly from the effective Schrödinger problems. Solve the Virasoro constraint that does not contain the mixed derivative. Then, show that the other one is automatically satisfied. You just specified a wide class of string solutions on the two-dimensional sphere.

5. Let us now study some basic properties of the solutions that we found. First express them in spherical coordinates. Show that the string is rigidly rotating with constant angular velocity. Specify the latter.

6. It would be nice that the strings we are describing are actually finite. Find a condition that must be obeyed so that the string is a finite, closed string. Do not forget that the physical time X^0 is not proportional to the time-like world-sheet coordinate x^0, but it is given by $X^0 = m_+ x^+ + m_- x^-$. In order to find the appropriate periodicity condition you need to freeze the physical time. It would be simpler to perform a boost of the world-sheet coordinates x^0 and x^1 to new ones σ^0 and σ^1, so the physical time is proportional to σ^0.

7. Freeze the *physical* time X^0 and study the derivative $\frac{d\theta}{d\sigma^1}$. Is it continuous? How do the points where this derivative is singular look like? What is their velocity? Do these points appear to all four classes of solutions? What is the value of the Pohlmeyer field at these points?

8. These solutions have some particularly interesting limits. How do these solutions look like when the pendulum solution tends to become the vacuum? These limits are the so called Berenstein-Maldacena-Nastase particle [7] and the giant hoop.

9. How do these solutions look like when the pendulum solution tends to become the "kink" solution that lies exactly between the oscillating and rotating solutions? These limits are the giant magnon [8] and single spike [9] limits.

10. Can you find another interesting limit where the solution looks like a rotating arc of a great circle? These are the Gubser-Klebanov-Polyakov strings [10].

All the limits of steps 8 to 10 were constructed one-by-one via educated ansatz guesses. It is amazing that with the appropriate use of the Weierstrass elliptic function and some knowledge on Pohlmeyer reduction you managed to construct them all at once as members of a much wider family, and furthermore you managed to describe them with very simple expressions. If you meet an obstacle that you cannot overcome at any of the steps of the problem, you may look for hints in [11].

6.3 Studying the Stability of Classical String Solutions

In the previous problem you constructed a very wide family of classical string solutions. These find applications in string theory and especially in the field of the holographic duality. For these applications, it would be nice to know whether these solutions are stable or unstable configurations. The classical treatment is to introduce small perturbations around these solutions and study their time evolution. In the case of strings on the two-dimensional sphere, we can be a little smarter than that. As we discussed in the previous problem, the equations of motion turn out to be equivalent to the sine-Gordon equation. Therefore, it suffices to study the stability of the corresponding solutions of the sine-Gordon equation in order to conclude whether the corresponding string solutions are stable or not. This is a much simpler task, as the sine-Gordon equation contains a single degree of freedom.

1. Go back to the first problem and write down the solutions of the sine-Gordon equation that depend solely on time or position.

2. Introduce a small perturbation to these solutions. Substitute them in the sine-Gordon equation and find the linearised equations for the small perturbations that you introduced. These equations can be solved via separation of variables. You will meet a well-known friend from Chap. 5, the $n = 1$ Lamé equation.

3. Recall that the unperturbed sine-Gordon solution you have in hand describes a finite closed string, and thus, it obeys an appropriate periodicity condition that you specified in the previous problem. This periodicity condition must be obeyed by the perturbation, too. Recalling the quasi-periodicity properties of the eigenfunctions of the Lamé problem, express the appropriate periodicity conditions for the small perturbations in terms of the parameters of the eigenfunctions of the Lamé problem. Review the previous problem and recall that the physical time X^0 is not proportional to the time-like world-sheet coordinate x^0, but it is given by $X^0 = m_+ x^+ + m_- x^-$. In order to express correctly the appropriate periodicity condition for the perturbation, you need to freeze the physical time. Perform a boost of the world-sheet coordinates to new ones σ^0 and σ^1, so the physical time is proportional to σ^0.

4. Find the time evolution of the small perturbations using the quasi-periodicity properties of the eigenfunctions of the Lamé problem. Are there perturbations with appropriate periodicity conditions that lead to instabilities?

5. Go back to the first problem. Compare the stability-instability of a string solution with the possibility of propagation of a superluminal kink on top of the corresponding sine-Gordon equation solution with velocity equal to the inverse of the velocity of the boost that connects the world-sheet coordinates σ^0 and σ^1 to x^0 and x^1. This is not a coincidence; such kink solutions are the full non-linear realization of the instability. Actually, you have already built the Pohlmeyer counterpart of this full non-linear string solution, which exposes the instability, in the first problem of this chapter.

If you meet an obstacle that you cannot overcome at any of the steps of the problem, you may look for hints in [12].

References

1. R. Rajaraman, *Solitons and Instantons: An Introduction to Solitons and Instantons in Quantum Field Theory*, North-Holland Personal Library. 15. North-Holland, pp. 34–45 (1989). ISBN 978-0-444-87047-6
2. J. Cuevas-Maraver, P.G. Kevrekidis, F. Williams, *The sine-Gordon Model and Its Applications: From Pendula and Josephson Junctions to Gravity and High-Energy Physics* (2014). ISBN 978-3-319-06721-6
3. D. Katsinis, I. Mitsoulas, G. Pastras, Dressed elliptic string solutions on $\mathbb{R} \times S^2$. Eur. Phys. J. C **78**(8), 668 (2018), arXiv:1806.07730 [hep-th]
4. D. Katsinis, I. Mitsoulas, G. Pastras, Salient features of dressed elliptic string solutions on $\mathbb{R} \times S^2$. Eur. Phys. J. C **79**(10), 869 (2019), arXiv:1903.01408 [hep-th]
5. J. Polchinski, *String Theory, Vol. 1: Introduction to the Bosonic String*, Cambridge Monographs on Mathematical Physics (Cambridge University Press, 2005). ISBN 978-0521672276

6. K. Pohlmeyer, Integrable Hamiltonian systems and interactions through quadratic constraints. Commun. Math. Phys. **46**, 207 (1976)
7. D.E. Berenstein, J.M. Maldacena, H.S. Nastase, Strings in Flat Space and pp Waves from $N = 4$ SuperYang-Mills. JHEP **0204**, 013 (2002), arXiv:hep-th/0202021
8. D.M. Hofman, J.M. Maldacena, Giant Magnons. J. Phys. A **39**, 13095 (2006), arXiv:hep-th/0604135
9. R. Ishizeki, M. Kruczenski, Single spike solutions for strings on S^2 and S^3. Phys. Rev. D **76**, 126006 (2007), arXiv:0705.2429 [hep-th]
10. S.S. Gubser, I.R. Klebanov, A.M. Polyakov, A semiclassical limit of the Gauge/string correspondence. Nucl. Phys. B **636**, 99 (2002), arXiv:hep-th/0204051
11. D. Katsinis, I. Mitsoulas, G. Pastras, Elliptic string solutions on $\mathbb{R} \times S^2$ and their pohlmeyer reduction. Eur. Phys. J. C **78**(11), 977 (2018), arXiv:1805.09301 [hep-th]
12. D. Katsinis, I. Mitsoulas, G. Pastras, Stability analysis of classical string solutions and the dressing method. JHEP **09**, 106 (2019), arXiv:1903.01412 [hep-th]

Solutions

Problems of Chap. 1

1.1 We may write the integral formula (1.33) in terms of the roots of the cubic polynomial as,

$$\pm z \sim \int_{\wp(z)}^{\infty} \frac{1}{\sqrt{4t^3 - g_2 t - g_3}} dt = \int_{\wp(z)}^{\infty} \frac{1}{\sqrt{4(t - e_1)(t - e_2)(t - e_3)}} dt. \quad (A.1)$$

It is obvious that the integrand has branch cuts on the complex plane with endpoints the three roots. We may select the branch cuts as the red lines in Fig. A.1.

Assume that there are three real roots. We may now apply the integral formula (A.1) using the right blue path of Fig. A.1 as the integration path. Let ω_1 be the half-period corresponding to the largest root. Then, we get

$$\omega_1 \sim \int_{e_1}^{+\infty} \frac{1}{\sqrt{4(t - e_1)(t - e_2)(t - e_3)}} dt. \quad (A.2)$$

Clearly, the integrand is everywhere real (and also positive). It vanishes at infinity as $t^{-3/2}$ and it diverges at the left boundary of the integration as $(t - e_1)^{-1/2}$, therefore the integral converges. Thus, we showed that the half-period associated with the largest root is congruent to a real number.

Similarly in the case of the three real roots, we may apply the integral formula (A.1) using the left blue path of Fig. A.1 as the integration path. Let ω_2 be the half-period corresponding to the largest root. Then, we get

© The Author(s), under exclusive license to Springer Nature Switzerland AG 2020
G. Pastras, *The Weierstrass Elliptic Function and Applications in Classical and Quantum Mechanics*, SpringerBriefs in Physics,
https://doi.org/10.1007/978-3-030-59385-8

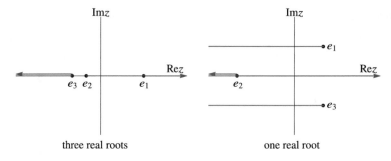

Fig. A.1 The branch cuts (red) and the integration paths (blue) used

$$
\begin{aligned}
\omega_2 &\sim \int_{e_3}^{-\infty} \frac{1}{\sqrt{4\,(t-e_1)\,(t-e_2)\,(t-e_3)}}\,dt \\
&= i \int_{-\infty}^{e_3} \frac{1}{\sqrt{4\,(e_1-t)\,(e_2-t)\,(e_3-t)}}\,dt.
\end{aligned}
\tag{A.3}
$$

The integrand is everywhere real and the integral converges for the same reasons as in previous case. Thus, the half-period associated with the smallest root is congruent to a purely imaginary number.

The situation is similar in the case of one real root. The only difference is that we may use the two different integration paths for the same root. Let ω_3 be the half-period corresponding to the real root. For the right blue path of Fig. A.1, we get

$$
\begin{aligned}
\omega_3 &\sim \int_{e_2}^{+\infty} \frac{1}{\sqrt{4\,(t-e_1)\,(t-e_2)\,(t-e_3)}}\,dt \\
&= \int_{e_2}^{+\infty} \frac{1}{\sqrt{4\left[(t-\operatorname{Re}e_1)^2+(\operatorname{Im}e_1)^2\right](t-e_2)}}\,dt,
\end{aligned}
\tag{A.4}
$$

whereas for the left blue path of Fig. A.1 we get

$$
\begin{aligned}
\omega_3 &\sim \int_{e_2}^{-\infty} \frac{1}{\sqrt{4\left[(t-\operatorname{Re}e_1)^2+(\operatorname{Im}e_1)^2\right](t-e_2)}}\,dt \\
&= i \int_{-\infty}^{e_2} \frac{1}{\sqrt{4\left[(t-\operatorname{Re}e_1)^2+(\operatorname{Im}e_1)^2\right](e_2-t)}}\,dt.
\end{aligned}
\tag{A.5}
$$

Both integrands are real and both integrals converge, implying that the half-period associated with the real root is congruent to both a real and a purely imaginary number.

1.2 Suppose that the two larger roots coincide, i.e. $e_1 = e_2 \equiv e_0$. Then, they are necessarily positive, as the three roots have to sum up to zero, implying that $e_3 = -2e_0$. In this case the Weierstrass differential equation reads,

$$\left(\frac{dy}{dz}\right)^2 = 4(y - e_0)^2 (y + 2e_0).$$

We perform the change of variable $y = e_0 + 3e_0/f^2$ and then the Weierstrass equation assumes the form,

$$f'^2 = 3e_0 \left(1 + f^2\right).$$

This one can be easily integrated to yield

$$f = \pm \sinh\left(\sqrt{3e_0}z + c\right).$$

Returning to the initial variable y, we find,

$$y = e_0 + \frac{3e_0}{\sinh^2\left(\sqrt{3e_0}z + c\right)}.$$

We know that by definition, the Weierstrass elliptic function should have a second order pole at $z = 0$. This sets the value of c to $in\pi$, $n \in \mathbb{Z}$, which implies that

$$\wp\left(z; 12e_0^2, -8e_0^3\right) = e_0 + \frac{3e_0}{\sinh^2\left(\sqrt{3e_0}z\right)}. \tag{A.6}$$

Therefore, in this case the Weierstrass elliptic function has degenerated to a simply periodic function. The only period is purely imaginary and it assumes the value $2\omega_2 = i\pi/\sqrt{3e_0}$.

Similarly, if the two smaller roots coincide, we have $e_2 = e_3 = -e_0$ and $e_1 = 2e_0$, where $e_0 > 0$. The Weierstrass differential equation reads,

$$\left(\frac{dy}{dz}\right)^2 = 4(y + e_0)^2 (y - 2e_0).$$

We perform the change of variable, $y = -e_0 + 3e_0/f^2$ and then the Weierstrass equation takes the form,

$$f'^2 = 3e_0 \left(1 - f^2\right).$$

This can be easily integrated to yield

$$f = \pm \sin\left(\sqrt{3e_0}z + c\right).$$

Returning to the initial variable y, we find,

$$y = -e_0 + \frac{3e_0}{\sin^2\left(\sqrt{3e_0}z + c\right)}.$$

Finally, requesting that the Weierstrass elliptic function has a second order pole at $z = 0$ sets the constant c to $n\pi$, $n \in \mathbb{Z}$ and we get,

$$\wp\left(z; 12e_0^2, 8e_0^3\right) = -e_0 + \frac{3e_0}{\sin^2\left(\sqrt{3e_0}z\right)}. \tag{A.7}$$

Similarly to the previous case, in this double root limit, the Weierstrass elliptic function has degenerated to a simply periodic function. Its only period is real and assumes the value $2\omega_1 = \pi/\sqrt{3e_0}$.

Finally, if all three roots coincide, they must be equal to zero, since their sum should vanish. In this case the Weierstrass differential equation reads,

$$\left(\frac{dy}{dz}\right)^2 = 4y^3.$$

This can trivially be integrated to acquire

$$y = \frac{1}{(z + c)^2}.$$

Once again, we demand that the Weierstrass elliptic function has a double pole at $z = 0$. Therefore $c = 0$ and

$$\wp\left(z; 0, 0\right) = \frac{1}{z^2}. \tag{A.8}$$

To sum up, we found that whenever two roots coincide, the Weierstrass elliptic function degenerates to a function that is not doubly but simply periodic. If all three roots coincide, the Weierstrass elliptic function degenerates to a function that is not periodic at all.

1.3 Employing the definition (1.18) of the Weierstrass elliptic function, we have

$$\frac{1}{\mu^2}\wp\left(z; \omega_1, \omega_2\right) = \frac{1}{(\mu z)^2}$$
$$+ \sum_{\{m,n\}\neq\{0,0\}} \left(\frac{1}{(\mu z + 2m\mu\omega_1 + 2n\mu\omega_2)^2} - \frac{1}{(2m\mu\omega_1 + 2n\mu\omega_2)^2}\right),$$

or

$$\frac{1}{\mu^2}\wp\left(z; \omega_1, \omega_2\right) = \wp\left(\mu z; \mu\omega_1, \mu\omega_2\right). \tag{A.9}$$

Then, the definitions of the moduli g_2 and g_3, (1.24) and (1.25), imply that

$$g_2 \left(\mu \omega_1, \mu \omega_2 \right) = 60 \sum_{\{m,n\} \neq \{0,0\}} \frac{1}{(2m\mu\omega_1 + 2n\mu\omega_2)^4} = \frac{1}{\mu^4} g_2 \left(\omega_1, \omega_2 \right),$$

$$g_3 \left(\mu \omega_1, \mu \omega_2 \right) = 140 \sum_{\{m,n\} \neq \{0,0\}} \frac{1}{(2m\mu\omega_1 + 2n\mu\omega_2)^6} = \frac{1}{\mu^6} g_3 \left(\omega_1, \omega_2 \right).$$

As a direct consequence of the above, we have

$$\wp \left(z; g_2, g_3 \right) = \mu^2 \wp \left(\mu z; \frac{g_2}{\mu^4}, \frac{g_3}{\mu^6} \right), \tag{A.10}$$

which is the desired homogeneity property of the Weierstrass elliptic function.

Problems of Chap. 2

2.1 Integrating the homogeneity property of Weierstrass elliptic function (1.41) and using the definition (2.1) of the function ζ, we get

$$\zeta \left(z; g_2, g_3 \right) = \mu \zeta \left(\mu z; \frac{g_2}{\mu^4}, \frac{g_3}{\mu^6} \right) + c.$$

The defining condition (2.2), i.e. the fact that ζ is defined as an odd function, implies that $c = 0$, and, thus,

$$\zeta \left(z; g_2, g_3 \right) = \mu \zeta \left(\mu z; \frac{g_2}{\mu^4}, \frac{g_3}{\mu^6} \right). \tag{A.11}$$

Integrating once more and using the definition (2.8) we get

$$\sigma \left(z; g_2, g_3 \right) = c\sigma \left(\mu z; \frac{g_2}{\mu^4}, \frac{g_3}{\mu^6} \right).$$

The defining condition (2.9) implies that $c = 1/\mu$. Therefore,

$$\sigma \left(z; g_2, g_3 \right) = \frac{1}{\mu} \sigma \left(\mu z; \frac{g_2}{\mu^4}, \frac{g_3}{\mu^6} \right). \tag{A.12}$$

2.2 Let us first consider the case that the two larger roots coincide to the value e_0. In this case, the Weierstrass elliptic function is expressed in terms of hyperbolic functions as shown in Eq. (1.43). Then, the definition of the Weierstrass zeta function (2.1) implies that

$$\zeta\left(z;12e_0^2,-8e_0^3\right) = -\int dz \wp\left(z;12e_0^2,-8e_0^3\right)$$

$$= -\int dz \left(e_0 + \frac{3e_0}{\sinh^2\left(\sqrt{3e_0}z\right)}\right)$$

$$= -e_0 z + \sqrt{3e_0}\coth\left(\sqrt{3e_0}z\right) + c.$$

The Laurent series of the function $\sqrt{3e_0}\coth\left(\sqrt{3e_0}z\right)$ around $z = 0$ is

$$\sqrt{3e_0}\coth\left(\sqrt{3e_0}z\right) = \frac{1}{z} + \mathcal{O}(z).$$

Therefore, the defining condition (2.2) implies that the integration constant c vanishes. Thus,

$$\zeta\left(z;12e_0^2,-8e_0^3\right) = -e_0 z + \sqrt{3e_0}\coth\left(\sqrt{3e_0}z\right). \tag{A.13}$$

In a similar manner, the definition of the Weierstrass sigma function (2.8) implies that

$$\ln\sigma\left(z;12e_0^2,-8e_0^3\right) = \int dz \zeta\left(z;12e_0^2,-8e_0^3\right)$$

$$= \int\left(-e_0 z + \sqrt{3e_0}\coth\left(\sqrt{3e_0}z\right)\right)$$

$$= -\frac{1}{2}e_0 z^2 + \ln\sinh\left(\sqrt{3e_0}z\right) + c,$$

or

$$\sigma\left(z;12e_0^2,-8e_0^3\right) = e^c \sinh\left(\sqrt{3e_0}z\right)e^{-\frac{1}{2}e_0 z^2}.$$

Taylor expanding the above around $z = 0$ yields

$$\sigma\left(z;12e_0^2,-8e_0^3\right) = e^c\sqrt{3e_0}z + \mathcal{O}\left(z^2\right).$$

Therefore, the defining condition (2.9) implies that the appropriate value for the integration constant is given by $e^c = 1/\sqrt{3e_0}$, and, thus,

$$\sigma\left(z;12e_0^2,-8e_0^3\right) = \frac{\sinh\left(\sqrt{3e_0}z\right)}{\sqrt{3e_0}}e^{-\frac{1}{2}e_0 z^2}. \tag{A.14}$$

It is trivial to repeat the above for the case that the two smaller roots coincide to the value $-e_0$ and find

$$\zeta \left(z; 12e_0^2, 8e_0^3\right) = e_0 z + \sqrt{3e_0} \cot\left(\sqrt{3e_0} z\right),$$ (A.15)

$$\sigma \left(z; 12e_0^2, 8e_0^3\right) = \frac{\sin\left(\sqrt{3e_0} z\right)}{\sqrt{3e_0}} e^{\frac{1}{2}e_0 z^2}.$$ (A.16)

2.3 The defining Eqs. (2.1) and (2.2) imply that

$$\zeta \left(z\right) = \frac{1}{z} - \int_0^z \left(\wp\left(w\right) - \frac{1}{w^2}\right) dw.$$ (A.17)

Therefore,

$$\zeta \left(-z\right) = -\frac{1}{z} - \int_0^{-z} \left(\wp\left(w\right) - \frac{1}{w^2}\right) dw$$

$$= -\frac{1}{z} + \int_0^z \left(\wp\left(-w'\right) - \frac{1}{(-w')^2}\right) dw'$$

$$= -\frac{1}{z} + \int_0^z \left(\wp\left(w'\right) - \frac{1}{w'^2}\right) dw',$$

where we made the change of the integration variable $w = -w'$ and we used the fact that $\wp$ is an even function. The latter implies that

$$\zeta \left(-z\right) = -\zeta \left(z\right),$$ (A.18)

meaning that ζ is an odd function.

Similarly, Eqs. (2.8) and (2.9) imply

$$\sigma \left(z\right) = z e^{\int_0^z \left(\zeta(w) - \frac{1}{w}\right) dw}.$$ (A.19)

Therefore,

$$\sigma \left(-z\right) = -z e^{\int_0^{-z} \left(\zeta(w) - \frac{1}{w}\right) dw} = -z e^{-\int_0^z \left(\zeta(-w') - \frac{1}{-w'}\right) dw'}$$

$$= -z e^{\int_0^z \left(\zeta(w') - \frac{1}{w'}\right) dw'}$$

or else

$$\sigma \left(-z\right) = -\sigma \left(z\right),$$ (A.20)

meaning that σ is an odd function.

2.4 First, we will prove the quasi-periodicity property

$$\zeta \left(z + 2m\omega_1 + 2n\omega_2\right) = \zeta \left(z\right) + 2m\zeta \left(\omega_1\right) + 2n\zeta \left(\omega_2\right)$$ (A.21)

for $n = 0$, namely,

$$\zeta \left(z + 2m\omega_1\right) = \zeta \left(z\right) + 2m\zeta \left(\omega_1\right).$$ (A.22)

This trivially holds for $m = 0$. Assuming that it holds for a given value m, we will show that it does hold for $m + 1$. Using property (2.5), we have

$$\zeta\left(z + 2\left(m + 1\right)\omega_1\right) = \zeta\left(z + 2m\omega_1\right) + 2\zeta\left(\omega_1\right)$$
$$= \zeta\left(z\right) + 2m\zeta\left(\omega_1\right) + 2\zeta\left(\omega_1\right)$$
$$= \zeta\left(z\right) + 2\left(m + 1\right)\zeta\left(\omega_1\right).$$

Similarly, one can show that (A.22) holds for $m - 1$. Thus, Eq. (A.22) holds for any m by induction. Therefore, the quasi-periodicity property (A.21) holds for $n = 0$. Assuming that it holds for a given value n, we will show that it holds for $n + 1$. Once again, using property (2.5), we have

$$\zeta\left(z + 2m\omega_1 + 2\left(n + 1\right)\omega_2\right) = \zeta\left(z + 2m\omega_1 + 2n\omega_2\right) + 2\zeta\left(\omega_2\right)$$
$$= \zeta\left(z\right) + 2m\zeta\left(\omega_1\right) + 2n\zeta\left(\omega_2\right) + 2\zeta\left(\omega_2\right)$$
$$= \zeta\left(z\right) + 2m\zeta\left(\omega_1\right) + 2\left(n + 1\right)\zeta\left(\omega_2\right).$$

Similarly, one can show that (A.21) holds for $n - 1$. Therefore, we have shown that (A.21) holds for any m and n by induction.

In a similar manner, we proceed to prove the quasi-periodicity property

$$\sigma\left(z + 2m\omega_1 + 2n\omega_2\right)\left(-1\right)^{m+n+mn}e^{\left(2m\zeta\left(\omega_1\right)+2n\zeta\left(\omega_2\right)\right)\left(z+m\omega_1+n\omega_2\right)}\sigma\left(z\right) \quad \text{(A.23)}$$

for $n = 0$, namely,

$$\sigma\left(z + 2m\omega_1\right) = \left(-1\right)^m e^{2m\zeta\left(\omega_1\right)\left(z+m\omega_1\right)}\sigma\left(z\right). \quad \text{(A.24)}$$

This property trivially holds for $m = 0$. Assuming that it holds for a given value of m, we will show that it does hold for $m + 1$. Employing property (2.12), we have

$$\sigma\left(z + 2\left(m + 1\right)\omega_1\right) = -e^{2\zeta\left(\omega_1\right)\left(z+2m\omega_1+m\omega_1\right)}\sigma\left(z + 2m\omega_1\right)$$
$$= -e^{2\zeta\left(\omega_1\right)\left(z+\left(2m+1\right)\omega_1\right)}\left(-1\right)^m e^{2m\zeta\left(\omega_1\right)\left(z+m\omega_1\right)}\sigma\left(z\right)$$
$$= \left(-1\right)^{m+1}e^{2\zeta\left(\omega_1\right)\left(\left(m+1\right)z+\left(m^2+2m+1\right)\omega_1\right)}\sigma\left(z\right)$$
$$= \left(-1\right)^{\left(m+1\right)}e^{2\left(m+1\right)\zeta\left(\omega_1\right)\left(z+\left(m+1\right)\omega_1\right)}\sigma\left(z\right).$$

Similarly one can show that (A.24) holds for $m - 1$. Therefore, property (A.24) holds for any m by induction. This implies that the quasi periodicity property (A.23) holds for $n = 0$. Assuming that it does hold for a given value of n, we will show that it does hold for $n + 1$,

$$\sigma\left(z+2m\omega_1+2\left(n+1\right)\omega_2\right)=-e^{2\zeta\left(\omega_2\right)\left(z+2m\omega_1+2n\omega_2+\omega_2\right)}\sigma\left(z+2m\omega_1+2n\omega_2\right)$$

$$=-e^{2\zeta\left(\omega_2\right)\left(z+2m\omega_1+2n\omega_2+\omega_2\right)}\left(-1\right)^{m+n+mn}e^{\left(2m\zeta\left(\omega_1\right)+2n\zeta\left(\omega_2\right)\right)\left(z+m\omega_1+n\omega_2\right)}\sigma\left(z\right)$$

$$=\left(-1\right)^{m+n+1+mn}e^{2m\zeta\left(\omega_1\right)\left(z+m\omega_1+n\omega_2\right)}e^{2\zeta\left(\omega_2\right)\left(\left(n+1\right)z+\left(mn+2m\right)\omega_1+\left(n^2+2n+1\right)\omega_2\right)}\sigma\left(z\right)$$

$$=\left(-1\right)^{m+n+1+mn}e^{2m\zeta\left(\omega_1\right)\left(z+m\omega_1+\left(n+1\right)\omega_2\right)}e^{2\left(n+1\right)\zeta\left(\omega_2\right)\left(z+m\omega_1+\left(n+1\right)\omega_2\right)}$$

$$\times\, e^{m\left(2\zeta\left(\omega_2\right)\omega_1-2\zeta\left(\omega_1\right)\omega_2\right)}\sigma\left(z\right)$$

$$=\left(-1\right)^{m+n+1+mn}e^{\left(2m\zeta\left(\omega_1\right)+2\left(n+1\right)\zeta\left(\omega_2\right)\right)\left(z+m\omega_1+\left(n+1\right)\omega_2\right)}e^{-i\pi m}\sigma\left(z\right)$$

$$=\left(-1\right)^{m+\left(n+1\right)+m\left(n+1\right)}e^{\left(2m\zeta\left(\omega_1\right)+2\left(n+1\right)\zeta\left(\omega_2\right)\right)\left(z+m\omega_1+\left(n+1\right)\omega_2\right)}\sigma\left(z\right),$$

where in the last step we used property (2.7). Similarly, one can show that (A.23) holds for $n-1$, and, thus, for any m and n.

2.5 The addition formula (2.18) implies that

$$\wp\left(z+\omega_1\right)=-\wp\left(z\right)-\wp\left(\omega_1\right)+\left(\frac{1}{2}\frac{\wp'\left(z\right)-\wp'\left(\omega_1\right)}{\wp\left(z\right)-\wp\left(\omega_1\right)}\right)^2.$$

Since $\wp\left(\omega_1\right)=e_1$ and $\wp$ is stationary at all half-periods, it follows that

$$\wp\left(z+\omega_1\right)=-\wp\left(z\right)-e_1+\left(\frac{1}{2}\frac{\wp'\left(z\right)}{\wp\left(z\right)-e_1}\right)^2.$$

The Weierstrass differential equation (1.29) can be used in order to substitute $\wp'\left(z\right)^2$ in the above expression. When written in terms of the roots, the Weierstrass equation assumes the form $\wp'\left(z\right)^2=4\left(\wp\left(z\right)-e_1\right)\left(\wp\left(z\right)-e_2\right)\left(\wp\left(z\right)-e_3\right)$, and, thus,

$$\wp\left(z+\omega_1\right)=-\wp\left(z\right)-e_1+\frac{\left(\wp\left(z\right)-e_2\right)\left(\wp\left(z\right)-e_3\right)}{\wp\left(z\right)-e_1}.$$

Finally, using the fact that the three roots sum to zero, we find

$$\wp\left(z+\omega_1\right)=\frac{e_1\wp\left(z\right)+e_1^2+e_2e_3}{\wp\left(z\right)-e_1}=e_1+\frac{2e_1^2+e_2e_3}{\wp\left(z\right)-e_1}.\qquad\text{(A.25)}$$

Similarly, it can be shown that

$$\wp\left(z+\omega_2\right)=e_3+\frac{2e_3^2+e_1e_2}{\wp\left(z\right)-e_3},\quad\wp\left(z+\omega_3\right)=e_2+\frac{2e_2^2+e_3e_1}{\wp\left(z\right)-e_2}.\qquad\text{(A.26)}$$

2.6 We will write the elliptic function $\frac{1}{2}\frac{\wp'(z)-\wp'(w)}{\wp(z)-\wp(w)}$, being considered a function of z, in terms of the Weierstrass function ζ and its derivatives, taking advantage of the techniques analysed in Sect. 2.2. In order to do, so we need an irreducible set of poles of this function and the principal part of its Laurent series at each pole.

We have already shown in Sect. 1.1 that given a meromorphic function $f(z)$, the function f'/f has only first order poles at the positions f has a root or a pole with residues equal to the multiplicity of the root and the opposite of the multiplicity of the pole respectively. We will apply that for the function $\wp(z) - \wp(w)$. The latter has obviously only a second order pole in a position congruent to $z = 0$ in each cell. The function $\wp(z) - \wp(w)$ obviously vanishes at $z = w$ and $z = -w$, which are not congruent to each other in general. Therefore, since it is a second order elliptic function, in each cell it has only two first order roots congruent to $z = w$ and $z = -w$. Consequently, an irreducible set of poles of the function $\frac{\wp'(z)}{\wp(z)-\wp(w)}$ are $z = 0$, $z = w$ and $z = -w$ and the corresponding Laurent series read,

$$\frac{\wp'(z)}{\wp(z) - \wp(w)} = -\frac{2}{z} + O(z^0),$$

$$\frac{\wp'(z)}{\wp(z) - \wp(w)} = \frac{1}{z - w} + O\left((z - w)^0\right),$$

$$\frac{\wp'(z)}{\wp(z) - \wp(w)} = \frac{1}{z + w} + O\left((z + w)^0\right).$$

Similarly, the function $-\frac{\wp'(w)}{\wp(z)-\wp(w)}$ has poles only at the locations where the function $\wp(z) - \wp(w)$ has roots. We have already shown that an irreducible set of roots of the latter is $z = w$ and $z = -w$. The corresponding Laurent series read,

$$-\frac{\wp'(w)}{\wp(z) - \wp(w)} = -\frac{\wp'(w)}{\wp'(w)(z - w)} + O\left((z - w)^0\right) = -\frac{1}{z - w} + O\left((z - w)^0\right),$$

$$-\frac{\wp'(w)}{\wp(z) - \wp(w)} = -\frac{\wp'(w)}{\wp'(-w)(z + w)} + O\left((z + w)^0\right) = \frac{1}{z + w} + O\left((z + w)^0\right).$$

Combining the above information, it turns out that the function $\frac{1}{2}\frac{\wp'(z)-\wp'(w)}{\wp(z)-\wp(w)}$ has poles only at $z = 0$ and $z = -w$ and the corresponding Laurent series read,

$$\frac{1}{2}\frac{\wp'(z) - \wp'(w)}{\wp(z) - \wp(w)} = -\frac{1}{z} + O(z^0),$$

$$\frac{1}{2}\frac{\wp'(z) - \wp'(w)}{\wp(z) - \wp(w)} = \frac{1}{z + w} + O\left((z + w)^0\right).$$

The above provides the necessary information in order to write the elliptic function $\frac{1}{2}\frac{\wp'(z)-\wp'(w)}{\wp(z)-\wp(w)}$ in terms of function ζ and its derivatives. Applying formula (2.15), we get

$$\frac{1}{2}\frac{\wp'(z) - \wp'(w)}{\wp(z) - \wp(w)} = \zeta(z + w) - \zeta(z) + c(w).$$

Interchanging z and w, we get in a similar manner

$$\frac{1}{2} \frac{\wp'(z) - \wp'(w)}{\wp(z) - \wp(w)} = \zeta(z + w) - \zeta(w) + c(z).$$

The above two relations imply that

$$\frac{1}{2} \frac{\wp'(z) - \wp'(w)}{\wp(z) - \wp(w)} = \zeta(z + w) - \zeta(z) - \zeta(w) + c.$$

Finally, the fact that $\wp'$ and ζ are odd functions, whereas $\wp$ is an even function implies that $c = 0$, which results in the desired pseudo-addition formula

$$\frac{1}{2} \frac{\wp'(z) - \wp'(w)}{\wp(z) - \wp(w)} = \zeta(z + w) - \zeta(z) - \zeta(w). \tag{A.27}$$

2.7 We differentiate the pseudo-addition theorem for Weierstrass ζ function (2.20) with respect to z and w. We get

$$-\wp(z + w) + \wp(z) = \frac{1}{2} \frac{\wp''(z)(\wp(z) - \wp(w)) - \wp'(z)(\wp'(z) - \wp'(w))}{(\wp(z) - \wp(w))^2},$$

$$-\wp(z + w) + \wp(w) = -\frac{1}{2} \frac{\wp''(w)(\wp(z) - \wp(w)) - \wp'(w)(\wp'(z) - \wp'(w))}{(\wp(z) - \wp(w))^2}.$$

Adding the above two equations yields

$$-2\wp(z + w) + \wp(z) + \wp(w)$$

$$= \frac{1}{2} \frac{(\wp''(z) - \wp''(w))(\wp(z) - \wp(w)) - (\wp'(z) - \wp'(w))^2}{(\wp(z) - \wp(w))^2}$$

$$= \frac{1}{2} \frac{\wp''(z) - \wp''(w)}{\wp(z) - \wp(w)} - \frac{1}{2} \left(\frac{\wp'(z) - \wp'(w)}{\wp(z) - \wp(w)} \right)^2.$$

The second derivative of Weierstrass elliptic function can easily be expressed in terms of the Weierstrass elliptic function itself. Differentiating the Weierstrass differential equation, we get

$$\wp'' = 6\wp^2 - \frac{g_2}{2}, \tag{A.28}$$

which implies that

$$\wp''(z) - \wp''(w) = 6 \left(\wp^2(z) - \wp^2(w) \right).$$

Therefore, we conclude that

$$-2\wp(z + w) + \wp(z) + \wp(w) = 3 \frac{\wp^2(z) - \wp^2(w)}{\wp(z) - \wp(w)} - \frac{1}{2} \left(\frac{\wp'(z) - \wp'(w)}{\wp(z) - \wp(w)} \right)^2$$

and finally,

$$\wp(z+w) = -\wp(z) - \wp(w) - \frac{1}{4}\left(\frac{\wp'(z) - \wp'(w)}{\wp(z) - \wp(w)}\right)^2, \qquad (A.29)$$

which is the addition theorem for the Weierstrass elliptic function (2.18).

2.8 The addition formula for Weierstrass elliptic function states that

$$\wp(z+w) = -\wp(z) - \wp(w) + \left(\frac{1}{2}\frac{\wp'(z) - \wp'(w)}{\wp(z) - \wp(w)}\right)^2.$$

The duplication formula requires taking the limit $w \to z$. At this limit the fraction in the right hand side of the addition formula becomes indeterminate. It is simple though to use Hospital's rule to find

$$\wp(2z) = -2\wp(z) + \left(\frac{1}{2}\frac{\wp''(z)}{\wp'(z)}\right)^2. \qquad (A.30)$$

One can eliminate the second derivative of $\wp$. For this purpose we have to differentiate the Weierstrass equation to yield

$$\wp''(z) = 6\wp^2(z) - \frac{g_2}{2}. \qquad (A.31)$$

Then, the duplication formula assumes the form

$$\wp(2z) = -2\wp(z) + \frac{\left(6\wp^2(z) - \frac{g_2}{2}\right)^2}{4\left(4\wp^3(z) - g_2\wp(z) - g_3\right)}$$

or

$$\wp(2z) = \frac{1}{4}\wp(z) + \frac{3g_2\wp^2(z) + 9g_3\wp(z) + \frac{g_2^2}{4}}{4\wp'^2(z)}. \qquad (A.32)$$

Let's now derive the duplication formula expressing $\wp(2z)$ in terms of $\zeta(z)$ and its derivatives. Let $2\omega_1$ and $2\omega_2$ be the two fundamental periods of $\wp(z)$. Then the function $\wp(2z)$ is an elliptic function with fundamental periods equal to ω_1 and ω_2. This obviously means that it is also an elliptic function with periods $2\omega_1$ and $2\omega_2$, but in the parallelogram defined by the latter, there exist four fundamental period parallelograms of $\wp(2z)$. Therefore, the function $\wp(2z)$, as an elliptic function with periods $2\omega_1$ and $2\omega_2$ is an order 8 elliptic function with four double poles at positions congruent to $z = 0$, $z = \omega_1$, $z = \omega_2$ and, $z = \omega_3$. The principal part of the Laurent series of $\wp(2z)$ at each of these poles reads

$$\wp(2z) = \frac{1}{4}\frac{1}{(z - z_0)^2} + \mathcal{O}\left((z - z_0)^2\right).$$

We have the necessary data to express $\wp(2z)$ in terms of the function $\zeta(z)$ and its derivatives. Formula (2.15) implies that

$$\wp(2z) = C - \frac{1}{4}\zeta'(z) - \frac{1}{4}\zeta'(z - \omega_1) - \frac{1}{4}\zeta'(z - \omega_2) - \frac{1}{4}\zeta'(z - \omega_1 - \omega_2)$$

$$= C + \frac{1}{4}\wp(z) + \frac{1}{4}\wp(z - \omega_1) + \frac{1}{4}\wp(z - \omega_2) + \frac{1}{4}\wp(z - \omega_3).$$

Finding the Laurent series of the above equation at the region of $z = 0$, we get

$$\frac{1}{4z^4} + \mathcal{O}(z^2) = C + \frac{1}{4z^4} + e_1 + e_2 + e_3 + \mathcal{O}(z^2),$$

implying that $C = 0$, since the three roots e_1, e_2 and e_3 sum to zero. Using the results of problem 2.5, we get

$$\wp(2z) = \frac{1}{4}\wp(z) + \frac{1}{4}\left(e_1 + \frac{2e_1^2 + e_2 e_3}{\wp(z) - e_1}\right) + \frac{1}{4}\left(e_3 + \frac{2e_3^2 + e_1 e_2}{\wp(z) - e_3}\right) + \frac{1}{4}\left(e_2 + \frac{2e_2^2 + e_3 e_1}{\wp(z) - e_2}\right).$$

After some algebra and using the fact that the three roots sum to zero, we get

$$\wp(2z) = \frac{1}{4}\wp(z) + \frac{\left(2\left(e_1^2 + e_2^2 + e_3^2\right) + e_1 e_2 + e_2 e_3 + e_3 e_1\right)\wp^2(z)}{\wp'^2(z)}$$

$$+ \frac{\left(2\left(e_1^3 + e_2^3 + e_3^3\right) + 3e_1 e_2 e_3\right)\wp(z)}{\wp'^2(z)} + \frac{e_1^2 e_2^2 + e_2^2 e_3^2 + e_3^2 e_1^2}{\wp'^2(z)}.$$

The fact that the three roots sum to zero can be used to calculate the coefficients of the above expression. We find,

$$(e_1 + e_2 + e_3)^2 = \left(e_1^2 + e_2^2 + e_3^2\right) + 2\left(e_1 e_2 + e_2 e_3 + e_3 e_1\right) = 0,$$

which with the help of Eq. (1.39) implies that

$$e_1^2 + e_2^2 + e_3^2 = \frac{g_2}{2}. \tag{A.33}$$

Furthermore,

$$\left(e_1^2 + e_2^2 + e_3^2\right)(e_1 + e_2 + e_3)$$
$$= e_1^3 + e_2^3 + e_3^3 + \left(e_1^2 e_2 + e_2^2 e_3 + e_3^2 e_1 + e_1 e_2^2 + e_2 e_3^2 + e_3 e_1^2\right) = 0,$$

$$(e_1 + e_2 + e_3)^3$$
$$= \left(e_1^3 + e_2^3 + e_3^3\right) + 3\left(e_1^2 e_2 + e_2^2 e_3 + e_3^2 e_1 + e_1 e_2^2 + e_2 e_3^2 + e_3 e_1^2\right) + 6e_1 e_2 e_3 = 0.$$

Taking into account equation (1.40), the last two equations imply that

$$e_1^3 + e_2^3 + e_3^3 = 3e_1e_2e_3 = \frac{3g_3}{4}. \tag{A.34}$$

Finally,

$$(e_1e_2 + e_2e_3 + e_3e_1)^2 = e_1^2e_2^2 + e_2^2e_3^2 + e_3^2e_1^2 + 2e_1e_2e_3(e_1 + e_2 + e_3),$$

implying that

$$e_1^2e_2^2 + e_2^2e_3^2 + e_3^2e_1^2 = (e_1e_2 + e_2e_3 + e_3e_1)^2 = \frac{g_2^2}{16}. \tag{A.35}$$

Putting everything together, we get

$$\wp(2z) = \frac{1}{4}\wp(z) + \frac{3g_2\wp^2(z) + 9g_3\wp(z) + \frac{g_2^2}{4}}{4\wp'^2(z)}, \tag{A.36}$$

which is the desired duplication formula.

Problems of Chap. 4

4.1 When the cubic polynomial has a double root e_0, the third root necessarily equals $-2e_0$, as the three roots sum to zero. Therefore, the cubic polynomial in such a case equals,

$$Q(x) = 4(x - e_0)^2(x + 2e_0) = 4x^3 - 12e_0^2x + 8e_0^3. \tag{A.37}$$

Thus, a cubic polynomial with a double root obeys

$$g_2 = 12e_0^2, \quad g_3 = -8e_0^3, \tag{A.38}$$

or

$$g_2 > 0, \quad g_2^3 - 27g_3^2 = 0. \tag{A.39}$$

When $g_3 > 0$, the double root is negative, and, thus, the two smaller roots coincide, whereas when $g_3 < 0$, the double root is positive meaning that the two larger roots coincide.

In the case of the motion of a particle in a cubic potential, the solution is given by

$$x = \wp(t - t_0; -F_0, -E), \tag{A.40}$$
$$x = \wp(t - t_0 + \omega_2; -F_0, -E), \tag{A.41}$$

where the second solution exists only when there are three real roots. The moduli take the values $g_2 = -F_0$ and $g_3 = -E$ and therefore a double root exists when

$$F_0 < 0, \quad E = \pm\left(-\frac{F_0}{3}\right)^{\frac{3}{2}}. \tag{A.42}$$

These values correspond to the case the potential that possesses a local minimum and the energies equal the limiting values for the existence of a bounded motion.

When $E = +(-F_0/3)^{3/2}$, the double root equals $e_0 = (-F_0/12)^{1/2} > 0$. Therefore, the real period diverges, while the imaginary one assumes the value

$$2\omega_2 = i\pi\left(-\frac{3F_0}{4}\right)^{-\frac{1}{4}}. \tag{A.43}$$

Formula (1.43) suggests that in this case,

$$\wp\left(z; -F_0, -\left(-\frac{F_0}{3}\right)^{\frac{3}{2}}\right) = \left(-\frac{F_0}{12}\right)^{\frac{1}{2}} + \frac{\left(-\frac{3F_0}{4}\right)^{\frac{1}{2}}}{\sinh^2\left(\left(-\frac{3F_0}{4}\right)^{\frac{1}{4}}z\right)}, \tag{A.44}$$

implying that the unbounded and bounded motions degenerate to

$$x = \left(-\frac{F_0}{12}\right)^{\frac{1}{2}} + \frac{\left(-\frac{3F_0}{4}\right)^{\frac{1}{2}}}{\sinh^2\left(\left(-\frac{3F_0}{4}\right)^{\frac{1}{4}}(t - t_0)\right)}, \tag{A.45}$$

$$x = \left(-\frac{F_0}{12}\right)^{\frac{1}{2}} - \frac{\left(-\frac{3F_0}{4}\right)^{\frac{1}{2}}}{\cosh^2\left(\left(-\frac{3F_0}{4}\right)^{\frac{1}{4}}(t - t_0) + i\frac{\pi}{2}\right)}, \tag{A.46}$$

respectively. These motions describe a particle having exactly the energy of the local maximum, in the first case coming from the right and in the second coming from the left. The particle arrives at the position of the unstable equilibrium in infinite time.

When $E = -(-F_0/3)^{3/2}$ the double root equals $e_0 = -(-F_0/12)^{1/2} < 0$. The imaginary period diverges, while the real one assumes the value

$$2\omega_1 = \pi\left(-\frac{3F_0}{4}\right)^{-\frac{1}{4}}. \tag{A.47}$$

Formula (1.44) suggests that in this case,

$$\wp\left(z; -F_0, +\left(-\frac{F_0}{3}\right)^{\frac{3}{2}}\right) = -\left(-\frac{F_0}{12}\right)^{\frac{1}{2}} + \frac{\left(-\frac{3F_0}{4}\right)^{\frac{1}{2}}}{\sin^2\left(\left(-\frac{3F_0}{4}\right)^{\frac{1}{4}}z\right)}, \tag{A.48}$$

implying that the unbounded and bounded motions degenerate to

$$x = \left(-\frac{F_0}{12}\right)^{\frac{1}{2}} + \frac{\left(-\frac{3F_0}{4}\right)^{\frac{1}{2}}}{\sin^2\left(\left(-\frac{3F_0}{4}\right)^{\frac{1}{4}}(t - t_0)\right)}, \tag{A.49}$$

$$x = \left(-\frac{F_0}{12}\right)^{\frac{1}{2}} + \lim_{x \to \infty} \frac{\left(-\frac{3F_0}{4}\right)^{\frac{1}{2}}}{\sin^2\left(\left(-\frac{3F_0}{4}\right)^{\frac{1}{4}}(t - t_0) + ix\right)} = \left(-\frac{F_0}{12}\right)^{\frac{1}{2}}, \tag{A.50}$$

respectively. The form of the limit of the bounded motion is also physically expected. The bounded motion ranges between e_3 and e_2. Since these two roots coincide, the bounded motion necessarily degenerates to a constant. The unbounded motion describes a point particle having exactly the energy of the local minimum coming from the right and getting reflected by the potential barrier, while the bounded motion describes a point particle resting at the local minimum equilibrium position.

The period of motion in this case assumes a finite value. The Taylor series of the potential at the region of the local minimum is

$$V(x) = -\left(-\frac{F_0}{3}\right)^{\frac{3}{2}} + 2\sqrt{-3F_0}\left(x + \left(-\frac{F_0}{12}\right)^{\frac{1}{2}}\right)^2 - 4\left(x + \left(-\frac{F_0}{12}\right)^{\frac{1}{2}}\right)^3. \tag{A.51}$$

Since the mass of the point particle has been taken equal to 2, the period of the small oscillations at the region of the local minimum is equal to

$$T_{\text{small}} = \frac{2\pi}{(-12F_0)^{\frac{1}{4}}} = 2\omega_1, \tag{A.52}$$

as expected.

4.2 The cubic polynomial associated with the problem of a point particle with energy E is

$$Q(x) = 4x^3 + F_0 x + E.$$

Let e_1, e_2 and e_3 be its three roots, appropriately ordered, as described in Sect. 3.1. They obviously obey

$$Q(e_i) = 4e_i^3 + F_0 e_i + E = 0.$$

The cubic polynomial associated with the problem of a point particle with energy $-E$ is

$$R(x) = 4x^3 + F_0 x - E.$$

This polynomial obeys

$$R(-e_i) = -4e_i^3 - F_0 e_i - E = -Q(e_i) = 0. \tag{A.53}$$

In other words, the roots of the problem with inverted energy are just the opposite of the roots of the initial problem. It is obvious that the appropriate ordering of the roots of the inverted energy problem is

$$e_1 (-E) = -e_3 (E), \quad e_2 (-E) = -e_2 (E), \quad e_3 (-E) = -e_1 (E). \qquad \text{(A.54)}$$

In the case the three roots are real, the above is obvious, since $e_1 > e_2 > e_3 \Rightarrow -e_1 < -e_2 < -e_3$. In the case there is only one real root, it also holds: $e_2 (-E)$ has to be identified to $-e_2 (E)$, since it has to be real, while $e_1 (-E)$ has to be identified to $-e_3 (E)$, so that it has a positive imaginary part.

In the case of three real roots, we use formula (3.1) to find that

$$2\omega_1 (-E) = \int_{-e_3}^{+\infty} \frac{dt}{\sqrt{(t + e_1) (t + e_2) (t + e_3)}}$$

$$= \int_{-\infty}^{e_3} \frac{dt'}{\sqrt{(e_1 - t') (e_2 - t') (e_3 - t')}} \qquad \text{(A.55)}$$

$$= -2i\omega_2 (E) = |2\omega_2 (E)|,$$

where we defined $t' = -t$. Similarly, in the case of one real root, we use formula (3.3) to find

$$2\omega_1 (-E) + 2\omega_2 (-E) = \int_{-e_2}^{+\infty} \frac{dt}{\sqrt{(t + e_1) (t + e_2) (t + e_3)}}$$

$$= \int_{-\infty}^{e_2} \frac{dt'}{\sqrt{(t' - e_2) (e_2 - t') (t' - e_3)}} \qquad \text{(A.56)}$$

$$= -2i (2\omega_1 (E) - 2\omega_2 (E))$$

$$= |2\omega_1 (E) - 2\omega_2 (E)|.$$

This completes the proof that the absolute value of the imaginary period corresponds in all cases to the "time of flight" or period of oscillations for a point particle with opposite energy.

4.3 The conservation of energy in the cubic potential problem reads

$$\dot{x}^2 - 4x^3 - F_0 x = E$$

or in terms of the roots

$$\dot{x}^2 = 4 (x - e_1) (x - e_2) (x - e_3). \qquad \text{(A.57)}$$

We perform the change of variable $x = e_3 + \frac{(e_3 - e_1)(e_3 - e_2)}{y - e_3}$. It is a matter of simple algebra to find that

$$\dot{x} = -\frac{(e_3 - e_1)(e_3 - e_2)}{(y - e_3)^2}\dot{y},$$

$$x - e_1 = \frac{(e_3 - e_1)(y - e_2)}{y - e_3},$$

$$x - e_2 = \frac{(e_3 - e_2)(y - e_1)}{y - e_3},$$

$$x - e_3 = \frac{(e_3 - e_1)(e_3 - e_2)}{y - e_3}.$$

Thus, in terms of the new variable y, the conservation of energy is written as

$$\frac{(e_3 - e_1)^2(e_3 - e_2)^2}{(y - e_3)^4}\dot{y}^2$$

$$= 4\frac{(e_3 - e_1)(y - e_2)}{y - e_3}\frac{(e_3 - e_2)(y - e_1)}{y - e_3}\frac{(e_3 - e_1)(e_3 - e_2)}{y - e_3}$$

or

$$\dot{y}^2 = 4(y - e_1)(y - e_2)(y - e_3). \tag{A.58}$$

Therefore, the conservation of energy is invariant under this transformation. We could say that the problem has a $\mathbb{Z}_2$ symmetry, since performing the change of variable twice leads to the initial variable,

$$x \rightarrow e_3 + \frac{(e_3 - e_1)(e_3 - e_2)}{x - e_3} \rightarrow e_3 + \frac{(e_3 - e_1)(e_3 - e_2)}{e_3 + \frac{(e_3 - e_1)(e_3 - e_2)}{x - e_3} - e_3} = x. \tag{A.59}$$

It is trivial to show that

$$\lim_{y \to -\infty} x = e_3, \quad \lim_{y \to e_3^-} x = -\infty, \quad \lim_{y \to e_3^+} x = +\infty,$$

$$\lim_{y \to e_2} x = e_1, \quad \lim_{y \to e_1} x = e_2, \quad \lim_{y \to +\infty} x = e_3.$$

As a consequence, the following intervals of x are mapped to intervals of y as

$$(-\infty, e_3] \rightarrow (-\infty, e_3],$$
$$[e_3, e_2] \rightarrow [e_1, +\infty),$$
$$[e_2, e_1] \rightarrow [e_2, e_1],$$
$$[e_1, +\infty) \rightarrow [e_3, e_2].$$

That means that the bounded motion in one problem is mapped to the unbounded motion of the other. Since the problems are identical, the two solutions, oscillatory and scattering, are also identical,

$$x_{\text{oscillatory}}(t) = y_{\text{oscillatory}}(t) = f(t), \tag{A.60}$$

$$x_{\text{scattering}}(t) = y_{\text{scattering}}(t) = g(t), \tag{A.61}$$

which implies

$$f(t) = e_3 + \frac{(e_3 - e_1)(e_3 - e_2)}{g(t) - e_3}, \quad g(t) = e_3 + \frac{(e_3 - e_1)(e_3 - e_2)}{f(t) - e_3}. \tag{A.62}$$

We do not know the exact form of the oscillatory and scattering motions, however the above equations connect them in a specific way. Let T be the period of the oscillatory motion and furthermore let the initial condition for the bounded motion be $f(0) = e_3$. In an obvious manner, after half a period the particle lies at the other point where its velocity vanishes $f\left(\frac{T}{2}\right) = e_2$. The connection between the two motions implies that

$$f(0) = e_3, \quad g(0) \to +\infty,$$

$$f\left(\frac{T}{2}\right) = e_2, \quad g\left(\frac{T}{2}\right) = e_1,$$

$$f(T) = e_3, \quad g(T) \to +\infty.$$

Thus, it turns out that the period of the oscillatory motion and the "time of flight" of the scattering motion are equal.

4.4 It is evident in Fig. 4.4 that there are two energies where a double root appears, namely $E = \pm\omega^2$.

For $E = \omega^2$, the two larger roots coincide,

$$e_1 = e_2 = e_0 = \frac{\omega^2}{3}.$$

Following the outcome of problem 1.2, the real period of the Weierstrass elliptic function diverges, whereas the imaginary one assumes the value

$$2\omega_2 = i\frac{\pi}{\omega}$$

and the Weierstrass elliptic function is expressed in terms of hyperbolic functions as

$$2\wp\left(z; \frac{4\omega^4}{3}, -\frac{8\omega^6}{27}\right) = \frac{\omega^2}{3} + \frac{\omega^2}{\sinh^2\omega z}.$$

The pendulum solution is expressed in terms of the bounded real solution in the real domain of Weierstrass equation, namely the Weierstrass elliptic function on the real axis shifted by ω_2,

$$2\wp\left(t+\omega_2;\frac{4\omega^4}{3},-\frac{8\omega^6}{27}\right)=\frac{\omega^2}{3}+\frac{\omega^2}{\sinh^2\left(\omega t+i\frac{\pi}{2}\right)}=\frac{\omega^2}{3}-\frac{\omega^2}{\cosh^2\omega t}.$$

Finally, Eq. (4.16) implies that the degenerate solution is written as

$$\cos\theta=-1+\frac{2}{\cosh^2\omega t}. \tag{A.63}$$

For $E=-\omega^2$, the two smaller roots coincide.

$$e_2=e_3=-e_0=-\frac{\omega^2}{3}.$$

Therefore, the imaginary period diverges, while the real one assumes the value

$$2\omega_1=\frac{\pi}{\sqrt{3e_0}}=\frac{\pi}{\omega}.$$

As in problem 4.1, the bounded real solution in the real domain of Weierstrass equation degenerates to a constant being equal to the double root,

$$2\wp\left(t+\omega_2;\frac{4\omega^4}{3},\frac{8\omega^6}{27}\right)=-\frac{\omega^2}{3}. \tag{A.64}$$

Finally, Eq. (4.16) implies

$$\cos\theta=1. \tag{A.65}$$

This solution describes the pendulum lying at the stable equilibrium position. Considering this solution as the limit of the oscillatory pendulum motion when the amplitude of the oscillation goes to zero, we conclude that the limit of the period of the oscillatory motion at zero amplitude is

$$T_{\text{oscillating}}=4\omega_1=\frac{2\pi}{\omega}, \tag{A.66}$$

as expected.

4.5 In the case $E>\omega^2$, it holds that $e_1=x_1$, $e_2=x_2$ and $e_3=x_3$. Using the addition formula for the Weierstrass elliptic function yields (2.7)

$$x(-t)=\ln\left[\frac{2}{\omega^2}\left(2\wp\left(\omega_1/2-t\right)-\frac{E}{3}\right)\right]$$

$$=\ln\left[\frac{2}{\omega^2}\left(-2\wp\left(\omega_1\right)-2\wp\left(-\omega_1/2-t\right)+2\left(\frac{1}{2}\frac{\wp'\left(\omega_1\right)-\wp'\left(-\omega_1/2-t\right)}{\wp\left(\omega_1\right)-\wp\left(-\omega_1/2-t\right)}\right)^2-2e_1\right)\right]$$

$$=\ln\left[\frac{2}{\omega^2}\left(-4e_1-2\wp\left(\omega_1/2+t\right)+\frac{1}{2}\left(\frac{\wp'\left(\omega_1/2+t\right)}{e_1-\wp\left(\omega_1/2+t\right)}\right)^2\right)\right].$$

Then, we use the Weierstrass differential equation (1.29) to substitute the derivative of the Weierstrass function

$x(-t)$

$$= \ln\left[\frac{2}{\omega^2}\left(-4e_1 - 2\wp(\omega_1/2+t) + 2\frac{(\wp(\omega_1/2+t)-e_2)(\wp(\omega_1/2+t)-e_3)}{\wp(\omega_1/2+t)-e_1}\right)\right]$$

$$= \ln\left[\frac{2}{\omega^2}\left(2\frac{(-e_1-e_2-e_3)\wp(\omega_1/2+t)+2e_1^2+e_2e_3}{\wp(\omega_1/2+t)-e_1}\right)\right]$$

$$= \ln\left[\frac{2}{\omega^2}\left(2\frac{2e_1^2+e_2e_3}{\wp(\omega_1/2+t)-e_1}\right)\right].$$

Using the specific form of the roots of our problem, we find

$$2e_1^2 + e_2e_3 = 2x_1^2 + x_2x_3$$

$$= 2\left(\frac{E}{6}\right)^2 + \left(-\frac{E}{12} + \frac{1}{4}\sqrt{E^2-\omega^4}\right)\left(-\frac{E}{12} - \frac{1}{4}\sqrt{E^2-\omega^4}\right)$$

$$= \frac{E^2}{18} + \frac{E^2}{144} - \frac{E^2-\omega^4}{16} = \frac{\omega^4}{16},$$

which implies that finally, we get

$$x(-t) = \ln\left[\frac{\omega^2}{4(\wp(\omega_1/2+t)-e_1)}\right]$$

$$= -\ln\left[\frac{2}{\omega^2}\left(2\wp(\omega_1/2+t) - \frac{E}{3}\right)\right] = -x(t). \tag{A.67}$$

The case $E < \omega^2$ is identical with the permutation $e_1 \leftrightarrow e_2$, $\omega_1 \leftrightarrow \omega_3$.

The essential difference between the transmitting solutions and the reflecting solutions is that in the former case the time of flight equals the real half-period, while in the latter equals the whole real period. Therefore, in the case of reflecting solutions, the problem is much simpler, since,

$$x(-t) = \ln\left[\frac{2}{\omega^2}\left(2\wp(\omega_1-t) - \frac{E}{3}\right)\right]$$

$$= \ln\left[\frac{2}{\omega^2}\left(2\wp(-\omega_1+t) - \frac{E}{3}\right)\right] \tag{A.68}$$

$$= \ln\left[\frac{2}{\omega^2}\left(2\wp(\omega_1+t) - \frac{E}{3}\right)\right] = x(t),$$

as expected.

4.6 The extrema of motion can be read from Table 4.6. They are $\ln\frac{4(e_1-e_2)}{\omega^2}$ and $\ln\frac{4(e_1-e_3)}{\omega^2}$. It is a matter of simple algebra to show that

$$(e_1 - e_3)(e_1 - e_2) = e_1^2 - (e_2 + e_3)e_1 + e_2e_3 = 2e_1^2 + e_2e_3. \qquad (A.69)$$

In problem 4.5 we found that $2e_1^2 + e_2e_3 = \frac{\omega^4}{16}$. Therefore, $\ln\frac{4(e_1-e_2)}{\omega^2} = -\ln\frac{4(e_1-e_3)}{\omega^2}$ and consequently the extrema of motion are opposite.

The periodicity property $x(T - t) = -x(t)$ can be easily proved by repeating all the steps of problem 4.5 with the substitution $t \to -t$.

The double root limit, as one can easily see in Fig. 4.6 corresponds to $E = \omega^2$. In this limit, the two smaller roots coincide to the value $e_2 = e_3 = -\omega^2/12$. The oscillatory motion is described by the bounded solution. As shown in problem 4.1, in this limit, the bounded solution degenerates to a constant equal to the value of the double root. Consequently,

$$x(t) = \ln\left[-\frac{2}{\omega^2}\left(2\frac{\omega^2}{12} - \frac{\omega^2}{3}\right)\right] = \ln 1 = 0, \qquad (A.70)$$

which is indeed the equilibrium position. The real period degenerates to the value

$$T = 2\omega_1 = \frac{\pi}{\sqrt{-3e_2}} = \frac{2\pi}{\omega}, \qquad (A.71)$$

as expected.

Problems of Chap. 5

5.1 The special solution is

$$\begin{aligned}
\tilde{y}(x; \omega_{1,2,3}) &= \sqrt{\wp(x) - e_{1,3,2}}\left(\zeta(x + \omega_{1,2,3}) + e_{1,3,2}x\right) \\
&= y_\pm(x; \omega_{1,2,3})\left(\zeta(x + \omega_{1,2,3}) + e_{1,3,2}x\right).
\end{aligned} \qquad (A.72)$$

The first and second derivatives of $y_\pm(x; \omega_{1,2,3})$ are given by Eqs. (5.7) and (5.8). These equations for the solution modulus being equal to any of the half-periods read

$$\frac{dy_\pm(x; \omega_{1,2,3})}{dx} = \frac{1}{2}\frac{\wp'(x)}{\wp(x) - e_{1,3,2}}y_\pm(x; \omega_{1,2,3}), \qquad (A.73)$$

$$\frac{d^2y_\pm(x; \omega_{1,2,3})}{dx^2} = \left(2\wp(x) + e_{1,3,2}\right)y_\pm(x; \omega_{1,2,3}). \qquad (A.74)$$

Therefore, we may express the first and second derivatives of $\tilde{y}(x; \omega_{1,2,3})$ as,

$$\begin{aligned}
\frac{d\tilde{y}(x; \omega_{1,2,3})}{dx} = \frac{dy_\pm(x; \omega_{1,2,3})}{dx}&\left(\zeta(x + \omega_{1,2,3}) + e_{1,3,2}x\right) \\
&+ y_\pm(x; \omega_{1,2,3})\left(-\wp(x + \omega_{1,2,3}) + e_{1,3,2}\right)
\end{aligned}$$

and

$$\frac{d^2\tilde{y}(x;\omega_{1,2,3})}{dx^2} = \frac{d^2 y_\pm(x;\omega_{1,2,3})}{dx^2}\left(\zeta\left(x+\omega_{1,2,3}\right)+e_{1,3,2}x\right)$$

$$+2\frac{dy_\pm(x;\omega_{1,2,3})}{dx}\left(-\wp\left(x+\omega_{1,2,3}\right)+e_{1,3,2}\right)$$

$$-y_\pm(x;\omega_{1,2,3})\left(\wp'\left(x+\omega_{1,2,3}\right)\right)$$

$$= \left(2\wp(x)+e_{1,3,2}\right)\left(\zeta\left(x+\omega_{1,2,3}\right)+e_{1,3,2}x\right)y_\pm(x;\omega_{1,2,3})$$

$$-\frac{\wp'(x)}{\wp(x)-e_{1,3,2}}\left(\wp\left(x+\omega_{1,2,3}\right)-e_{1,3,2}\right)y_\pm(x;\omega_{1,2,3})$$

$$-\wp'\left(x+\omega_{1,2,3}\right)y_\pm(x;\omega_{1,2,3}).$$

(A.75)

The Weierstrass elliptic function, with an argument shifted by a half-period can be calculated with the use of the addition theorem. This has been done in problem 2.5,

$$\wp\left(x+\omega_{1,2,3}\right)=e_{1,3,2}+\frac{\left(e_{1,3,2}-e_{3,2,1}\right)\left(e_{1,3,2}-e_{2,1,3}\right)}{\wp(x)-e_{1,3,2}}.$$

(A.76)

It is a direct consequence that

$$\wp'\left(x+\omega_{1,2,3}\right)=-\frac{\left(e_{1,3,2}-e_{3,2,1}\right)\left(e_{1,3,2}-e_{2,1,3}\right)}{\left(\wp(x)-e_{1,3,2}\right)^2}\wp'(x).$$

(A.77)

Combining the above two relations, we find

$$\frac{\wp'(x)}{\wp(x)-e_{1,3,2}}\left(\wp\left(x+\omega_{1,2,3}\right)-e_{1,3,2}\right)+\wp'\left(x+\omega_{1,2,3}\right)$$

$$=\frac{\wp'(x)}{\wp(x)-e_{1,3,2}}\frac{\left(e_{1,3,2}-e_{3,2,1}\right)\left(e_{1,3,2}-e_{2,1,3}\right)}{\wp(x)-e_{1,3,2}}$$

$$-\frac{\left(e_{1,3,2}-e_{3,2,1}\right)\left(e_{1,3,2}-e_{2,1,3}\right)}{\left(\wp(x)-e_{1,3,2}\right)^2}\wp'(x)=0.$$

The latter implies that Eq. (A.75) is written as

$$\frac{d^2\tilde{y}(x;\omega_{1,2,3})}{dx^2}=\left(2\wp(x)+e_{1,3,2}\right)\left(\zeta\left(x+\omega_{1,2,3}\right)+e_{1,3,2}x\right)y_\pm(x;\omega_{1,2,3})$$

$$=\left(2\wp(x)+e_{1,3,2}\right)\tilde{y}(x;\omega_{1,2,3}),$$

(A.78)

meaning that indeed the functions $\tilde{y}(x;\omega_{1,2,3})$ are eigenfunctions of the $n=1$ Lamé problem with eigenvalues

$$\lambda=-e_{1,3,2}.$$

(A.79)

5.2 We apply the quasi-periodicity properties of the Weierstrass zeta and sigma functions (2.5) and (2.12) respectively on the definition of the function u (5.17) and we get

$$u_\pm(x+2\omega_1;a) = \frac{\sigma(x+2\omega_1\pm a)}{\sigma(x+2\omega_1)\sigma(a)} e^{\mp\frac{a\zeta(\omega_1)}{\omega_1}(x+2\omega_1)}$$

$$= \frac{-\sigma(x\pm a)\,e^{2\zeta(\omega_1)(x\pm a+\omega_1)}}{-\sigma(x)\,e^{2\zeta(\omega_1)(x+\omega_1)}\sigma(a)} e^{\mp\frac{a\zeta(\omega_1)}{\omega_1}(x+2\omega_1)}$$

$$= \frac{\sigma(x\pm a)}{\sigma(x)\sigma(a)} e^{\mp\frac{a\zeta(\omega_1)}{\omega_1}x} = u_\pm(x;a).$$

Thus, indeed, the functions $u_\pm(x;a)$ are periodic functions of x with the same period as the potential.

5.3 We follow the derivation for the unbounded potential. In the following, b is considered real. For a in the segment $[0, \omega_1]$, $a = b$. Then,

$$\overline{\psi_\pm(x;b)} = \frac{\sigma(x-\omega_2\pm b)\sigma(-\omega_2)}{\sigma(x-\omega_2)\sigma(-\omega_2\pm b)} e^{-\zeta(\pm b)x}$$

$$= \frac{\sigma(x+\omega_2\pm b)\,e^{-2\zeta(\omega_2)(x\pm b)}\sigma(\omega_2)e^{-\zeta(\pm b)x}}{\sigma(x+\omega_2)\,e^{-2\zeta(\omega_2)x}\sigma(+\omega_2\pm b)\,e^{-2\zeta(\omega_2)(\pm b)}}$$

$$= \psi_\pm(x;b).$$

For a in the segment $[0, \omega_2]$, $a = ib$,

$$\overline{\psi_\pm(x;ib)} = \frac{\sigma(x-\omega_2\mp ib)\sigma(-\omega_2)}{\sigma(x-\omega_2)\sigma(-\omega_2\mp ib)} e^{-\zeta(\mp ib)x}$$

$$= \frac{\sigma(x+\omega_2\mp ib)\,e^{-2\zeta(\omega_2)(x\mp ib)}\sigma(\omega_2)e^{-\zeta(\mp ib)x}}{\sigma(x+\omega_2)\,e^{-2\zeta(\omega_2)x}\sigma(\omega_2\mp ib)\,e^{-2\zeta(\omega_2)(\mp ib)}}$$

$$= \psi_\mp(x;ib).$$

For a in the segment $[\omega_2, \omega_3]$, $a = \omega_2 + b$,

$$\overline{\psi_\pm(x;\omega_2+b)} = \frac{\sigma(x-\omega_2\mp\omega_2\pm b)\sigma(-\omega_2)}{\sigma(x-\omega_2)\sigma(-\omega_2\mp\omega_2\pm b)} e^{-\zeta(\mp\omega_2\pm b)x}$$

$$= \frac{-\sigma(x+\omega_2\pm\omega_2\pm b)\,e^{-(2\pm2)\zeta(\omega_2)(x\pm b)}\sigma(\omega_2)e^{\pm2\zeta(\omega_2)x}e^{-\zeta(\pm\omega_2\pm b)x}}{-\sigma(x+\omega_2)\,e^{-2\zeta(\omega_2)x}\sigma(\omega_2\pm\omega_2\pm b)\,e^{-(2\pm2)\zeta(\omega_2)(\pm b)}}$$

$$= \psi_\pm(x;\omega_2+b).$$

Finally, for a in the segment $[\omega_1, \omega_3]$, $a = \omega_1 + ib$ and

$$\overline{\psi_\pm(x;\omega_1+ib)} = \frac{\sigma(x-\omega_2\pm\omega_1\mp ib)\sigma(-\omega_2)}{\sigma(x-\omega_2)\sigma(-\omega_2\pm\omega_1\mp ib)} e^{-\zeta(\pm\omega_1\mp ib)x}$$

$$= \frac{\sigma(x+\omega_2\mp\omega_1\mp ib)\,e^{2(\pm\zeta(\omega_1)-\zeta(\omega_2))(x\mp ib)}\sigma(\omega_2)e^{\mp2\zeta(\omega_1)x}e^{-\zeta(\mp\omega_1\mp ib)x}}{\sigma(x+\omega_2)\,e^{-2\zeta(\omega_2)x}\sigma(\omega_2\mp\omega_1\mp ib)\,e^{2(\pm\zeta(\omega_2)-\zeta(\omega_2))(\mp ib)}}$$

$$= \psi_\mp(x;\omega_1+ib),$$

which concludes the derivation of the reality properties of the eigenfunctions of the bounded $n = 1$ Lamé problem.

5.4 Following the derivation in the case of the unbounded potential presented in Sect. 5.1, we have

$$\frac{\sigma\,(x + 2\omega_1 + \omega_2 \pm a)\,\sigma\,(\omega_2)}{\sigma\,(x + 2\omega_1 + \omega_2)\,\sigma\,(\omega_2 \pm a)} = \frac{-e^{2\zeta(\omega_1)(x+\omega_1+\omega_2\pm a)}\sigma\,(x + \omega_2 \pm a)\,\sigma\,(\omega_2)}{-e^{2\zeta(\omega_1)(x+\omega_1+\omega_2)}\sigma\,(x + \omega_2)\,\sigma\,(\omega_2 \pm a)}$$

$$= e^{\pm 2a\zeta(\omega_1)}\frac{\sigma\,(x + \omega_2 \pm a)\,\sigma\,(\omega_2)}{\sigma\,(x + \omega_2)\,\sigma\,(\omega_2 \pm a)}.$$

Thus, we can write the eigenfunctions (5.41) of the bounded problem in the form of a Bloch wave

$$\psi_\pm\,(x; a) = u_\pm\,(x; a)\,e^{\pm ik(a)x}, \tag{A.80}$$

where

$$u_\pm\,(x; a) = \frac{\sigma\,(x + \omega_2 \pm a)\,\sigma\,(\omega_2)}{\sigma\,(x + \omega_2)\,\sigma\,(\omega_2 \pm a)}\,e^{\mp\frac{a\zeta(\omega_1)}{\omega_1}x}, \tag{A.81}$$

$$ik\,(a) = \frac{a\zeta\,(\omega_1) - \omega_1\zeta\,(a)}{\omega_1}, \tag{A.82}$$

with $u_\pm\,(x + 2\omega_1; a) = u_\pm\,(x; a)$. The function $k\,(a)$ is identical to the one in the unbounded potential. Therefore, the band structure of the bounded potential is identical to the band structure of the unbounded problem.

5.5 By direct computation starting from the expressions (5.6), we find

$$y_+y_- = \frac{\sigma\,(x + a)\,\sigma\,(x - a)}{\sigma^2\,(x)\,\sigma\,(a)\,\sigma\,(-a)}e^{-\zeta(a)x}e^{-\zeta(-a)x} = -\frac{\sigma\,(x + a)\,\sigma\,(x - a)}{\sigma^2\,(x)\,\sigma^2\,(a)}.$$

Using the pseudo-addition formula for the Weierstrass σ function (2.19), we get

$$y_+y_- = \wp\,(x) - \wp\,(a). \tag{A.83}$$

Similarly, using Eq. (5.7),

$$y_+'y_- - y_+y_-' = \left(\frac{1}{2}\frac{\wp'\,(x) - \wp'\,(a)}{\wp\,(x) - \wp\,(a)} - \frac{1}{2}\frac{\wp'\,(x) + \wp'\,(a)}{\wp\,(x) - \wp\,(a)}\right)y_+y_-$$

$$= -\frac{\wp'\,(a)}{\wp\,(x) - \wp\,(a)}\,(\wp\,(x) - \wp\,(a)) = -\wp'\,(a). \tag{A.84}$$

We repeat for the eigenfunctions (5.41) of the bounded $n = 1$ Lamé potential

$$\psi_+\psi_- = \frac{\sigma\,(x+\omega_2+a)\,\sigma\,(x+\omega_2-a)\,\sigma^2\,(\omega_2)}{\sigma^2\,(x+\omega_2)\,\sigma\,(\omega_2+a)\,\sigma\,(\omega_2-a)}\,e^{-\zeta(a)x}\,e^{-\zeta(-a)x}$$

$$= \frac{\sigma\,(x+\omega_2+a)\,\sigma\,(x+\omega_2-a)}{\sigma^2\,(x+\omega_2)\,\sigma^2\,(a)}\quad\frac{\sigma^2\,(\omega_2)\,\sigma^2\,(a)}{\sigma\,(\omega_2+a)\,\sigma\,(\omega_2-a)} \qquad\text{(A.85)}$$

$$= \frac{\wp\,(x+\omega_2) - \wp\,(a)}{e_3 - \wp\,(a)}$$

and

$$\psi_+{}'\psi_- - \psi_+\psi_-{}' = \left(\frac{1}{2}\frac{\wp'\,(x+\omega_2) - \wp'\,(a)}{\wp\,(x+\omega_2) - \wp\,(a)} - \frac{1}{2}\frac{\wp'\,(x+\omega_2) + \wp'\,(a)}{\wp\,(x+\omega_2) - \wp\,(a)}\right)\psi_+\psi_-$$

$$= -\frac{\wp'\,(a)}{\wp\,(x+\omega_2) - \wp\,(a)}\quad\frac{\wp\,(x+\omega_2) - \wp\,(a)}{e_3 - \wp\,(a)} = -\frac{\wp'\,(a)}{e_3 - \wp\,(a)}.$$
$$\text{(A.86)}$$

Therefore the "normalization" of the eigenfunctions of the bounded potential differs from the "normalization" of the eigenfunctions of the unbounded potential by a factor of $\sqrt{e_3 - \wp\,(a)}$. Of course this is just a constant and it could be included to the definition of the bounded eigenfunctions. However, such an inclusion would disturb the reality properties of the eigenfunctions, as the reality of this factor depends on the value of $\wp\,(a)$.

5.6 It is trivial that the square of the magnitude and the phase are given by

$$R^2 = y_+y_-,$$

$$\Phi = -\frac{i}{2}\ln\frac{y_+}{y_-}.$$

It follows that

$$\frac{d\Phi}{dx} = -\frac{i}{2}\frac{y_-\,y'_+y_- - y'_-y_+}{y_+}\,\frac{1}{y^2_-} = -\frac{i}{2}\frac{y'_+y_- - y'_-y_+}{y_+y_-}$$

Thus, as a direct consequence of Eqs. (5.51) and (5.52) we have

$$R^2\,(x;a) = \wp\,(x) - \wp\,(a)\,, \qquad\qquad\text{(A.87)}$$

$$\frac{d\Phi\,(x;a)}{dx} = \frac{i}{2}\frac{\wp'\,(a)}{\wp\,(x) - \wp\,(a)}. \qquad\qquad\text{(A.88)}$$

Indeed, both R^2 and $d\Phi/dx$ are elliptic functions of x.

Using the specific form of the eigenfunctions of the $n = 1$ Lamé problem (5.6), we find that the phase Φ is given by

$$\Phi(x; a) = -\frac{i}{2} \ln\left(\frac{\sigma(x+a)\sigma(-a)}{\sigma(x-a)\sigma(a)} e^{(\zeta(-a)-\zeta(a))x}\right)$$

$$= -\frac{i}{2}\left[\ln\left(-\frac{\sigma(x+a)}{\sigma(x-a)}\right) - 2\zeta(a)x\right].$$

(A.89)

We used the fact that the Weierstrass zeta and sigma functions are odd. Using the quasi-periodicity properties of the above functions, we get

$$\Phi(x + 2\omega_1; a) = -\frac{i}{2}\left[\ln\left(-\frac{\sigma(x+a+2\omega_1)}{\sigma(x-a+2\omega_1)}\right) - 2\zeta(a)(x+2\omega_1)\right]$$

$$= -\frac{i}{2}\left[\ln\left(-\frac{-\sigma(x+a)e^{2\zeta(\omega_1)(x+a+\omega_1)}}{-\sigma(x-a)e^{2\zeta(\omega_1)(x-a+\omega_1)}}\right) - 2\zeta(a)(x+2\omega_1)\right]$$

$$= -\frac{i}{2}\left[\ln\left(-\frac{\sigma(x+a)e^{2\zeta(\omega_1)(x+a+\omega_1)}}{\sigma(x-a)e^{2\zeta(\omega_1)(x-a+\omega_1)}}\right) - 2\zeta(a)x\right]$$

$$\qquad\qquad - 2i(\zeta(\omega_1)a - \zeta(a)\omega_1)$$

$$= \Phi(x; a) - 2i(\zeta(\omega_1)a - \zeta(a)\omega_1).$$

(A.90)

The above result is obviously compatible with the definition of the function k (5.18), since

$$\Phi(x + 2\omega_1; a) = \Phi(x; a) + 2\omega_1 k(a),$$

(A.91)

which implies that the eigenfunctions $y_\pm$ can be written as

$$y_\pm(x; a) = R(x; a) e^{\pm i(\Phi(x;a)-k(a))x} e^{\pm ik(a)x} \equiv u_\pm(x; a) e^{\pm ik(a)x},$$

(A.92)

where $u_\pm(x; a) = R(x; a) e^{\pm i(\Phi(x;a)-k(a))x}$ is obviously a periodic function with the same period as the potential.

5.7 Using the formula (5.7) and the definition of the superpotential (5.42), we find

$$A^\dagger y_\pm(x; a) = \left(-\frac{d}{dx} + W(x)\right)\frac{\sigma(x\pm a)}{\sigma(x)\sigma(\pm a)} e^{-\zeta(\pm a)x}$$

$$= -\frac{1}{2}\left(\frac{\wp'(x)\mp\wp'(a)}{\wp(x)-\wp(a)} - \frac{\wp'(x)}{\wp(x)-e_3}\right)\frac{\sigma(x\pm a)}{\sigma(x)\sigma(\pm a)} e^{-\zeta(\pm a)x}$$

$$\equiv f_\pm(x, a) y_\pm(x; a).$$

The Lamé eigenfunctions themselves are not elliptic functions. However, it is trivial that the prefactor $f_\pm(x, a)$ is an elliptic function of x, as well as an elliptic function of a.

Using the pseudo-addition formula for the Weierstrass ζ function (2.20), we find

$$f_\pm(x, a) = \zeta(x) + \zeta(\pm a) - \zeta(x\pm a) + \zeta(x+\omega_2) - \zeta(x) - \zeta(\omega_2)$$

$$= \zeta(\pm a) - \zeta(x\pm a) + \zeta(x+\omega_2) - \zeta(\omega_2).$$

Remembering that $f_\pm(x, a)$ as a function of x is an elliptic function, we may write it as a ratio of Weierstrass σ functions. $f_\pm(x, a)$ is clearly a second order elliptic function having two first order poles at $x = -\omega_2$ and $x = \mp a$. It also obviously has a zero at $x = 0$ since,

$$f_\pm(0, a) = \zeta(\pm a) - \zeta(\pm a) + \zeta(\omega_2) - \zeta(\omega_2) = 0.$$

Theorem 1.5 implies that the other zeros should be congruent to $x = \mp a - \omega_2$. Indeed,

$$f_\pm(-\omega_2 \mp a, a) = \zeta(\pm a) - \zeta(-\omega_2) + \zeta(\mp a) - \zeta(\omega_2) = 0.$$

The sum of the above poles equals the sum of the zeros, and, thus, we may write $f_\pm(x, a)$ as,

$$f_\pm(x, a) = C \frac{\sigma(x)\sigma(x + \omega_2 \pm a)}{\sigma(x + \omega_2)\sigma(x \pm a)}.$$

Requiring that $f_\pm(x, a)$ has residue equal to one at $x = -\omega_2$ yields $C = \frac{\sigma(-\omega_2 \pm a)}{\sigma(-\omega_2)\sigma(\pm a)}$, and, thus,

$$f_\pm(x, a) = \frac{\sigma(-\omega_2 \pm a)\sigma(x)\sigma(x + \omega_2 \pm a)}{\sigma(-\omega_2)\sigma(\pm a)\sigma(x + \omega_2)\sigma(x \pm a)}.$$

Therefore,

$$
\begin{aligned}
A^\dagger y_\pm(x; a) &= \frac{\sigma(-\omega_2 \pm a)\sigma(x)\sigma(x + \omega_2 \pm a)}{\sigma(-\omega_2)\sigma(\pm a)\sigma(x + \omega_2)\sigma(x \pm a)} \frac{\sigma(x \pm a)}{\sigma(x)\sigma(\pm a)} e^{-\zeta(\pm a)x} \\
&= \frac{\sigma(-\omega_2 \pm a)\sigma(x + \omega_2 \pm a)}{\sigma(-\omega_2)\sigma^2(\pm a)\sigma(x + \omega_2)} e^{-\zeta(\pm a)x} \\
&= -\frac{\sigma(\omega_2 \pm a)\sigma(-\omega_2 \pm a)}{\sigma^2(\omega_2)\sigma^2(\pm a)} \frac{\sigma(x + \omega_2 \pm a)\sigma(\omega_2)}{\sigma(x + \omega_2)\sigma(\omega_2 \pm a)} e^{-\zeta(\pm a)x} \\
&= (\wp(a) - e_3)\psi_\pm(x; a).
\end{aligned}
$$

$$(\text{A.93})$$

Indeed the action of the creation operator on the eigenfunctions of the unbounded $n = 1$ Lamé problem yields the eigenfunctions of the bounded problem multiplied with a constant. This constant is the same constant appearing in the "normalization" properties of the eigenfunctions in problem 5.5.

5.8 When the two larger roots coincide to the value e_0, the Weierstrass elliptic function degenerates to the form of Eq. (1.43). The real period diverges, whereas the imaginary period takes the value $2\omega_2 = i\pi/\sqrt{3e_0}$, as you showed in problem 1.2. Therefore the potential of the bounded $n = 1$ Lamé problem degenerates to the form

$$V(x) = 2\wp\left(z + i\frac{\pi}{2\sqrt{3e_0}}; 12e_0^2, -8e_0^3\right)$$

$$= 2e_0 + \frac{6e_0}{\sinh^2\left(\sqrt{3e_0}z + i\frac{\pi}{2}\right)} \quad \text{(A.94)}$$

$$= 2e_0 - \frac{6e_0}{\cosh^2\left(\sqrt{3e_0}z\right)}.$$

This is an attractive potential, which asymptotically tends to the constant value $2e_0$ at $\pm\infty$. Therefore, its spectrum definitely contains a continuous part corresponding to the energies that are larger than $2e_0$. But $2e_0 = -e_3$, thus this region of energies clearly corresponds to the infinite conduction band of the $n = 1$ Lamé potential. So, what happens to the finite valence band? This degenerates to a single value of energy $E = -e_0 = -e_1 = -e_2$. As the potential is attractive, in principle it may also contain bound states. Do the eigenstates of the finite valence band degenerate to a single bound and thus normalizable state?

In order to answer this question, we need to find the form of the eigenfunctions of the degenerate bound $n = 1$ Lamé problem. For this purpose we use the special form of the zeta and sigma functions in the double root limit that you calculated in problem 2.2, which are given by Eqs. (A.13) and (A.14). Using these formulae the eigenfunctions of the bounded $n = 1$ Lamé problem (5.41) assume the form

$$\psi_{\pm}(x; a) = \frac{\sigma(x + \omega_2 \pm a)\,\sigma(\omega_2)}{\sigma(x + \omega_2)\,\sigma(\omega_2 \pm a)}e^{-\zeta(\pm a)x},$$

$$= \frac{\frac{\sinh\left(\sqrt{3e_0}(x\pm a)+i\frac{\pi}{2}\right)}{\sqrt{3e_0}}e^{-\frac{1}{2}e_0\left(x\pm a+i\frac{\pi}{2\sqrt{3e_0}}\right)^2}\frac{\sinh\left(i\frac{\pi}{2}\right)}{\sqrt{3e_0}}e^{-\frac{1}{2}e_0\left(i\frac{\pi}{2\sqrt{3e_0}}\right)^2}}{\frac{\sinh\left(\sqrt{3e_0}(x)+i\frac{\pi}{2}\right)}{\sqrt{3e_0}}e^{-\frac{1}{2}e_0\left(x+i\frac{\pi}{2\sqrt{3e_0}}\right)^2}\frac{\sinh\left(\sqrt{3e_0}(\pm a)+i\frac{\pi}{2}\right)}{\sqrt{3e_0}}e^{-\frac{1}{2}e_0\left(\pm a+i\frac{\pi}{2\sqrt{3e_0}}\right)^2}}$$

$$\times\, e^{-\left(\mp e_0 a \pm \sqrt{3e_0}\coth\left(\sqrt{3e_0}a\right)\right)x}$$

$$= \frac{\cosh\left(\sqrt{3e_0}\,(x \pm a)\right)}{\cosh\left(\sqrt{3e_0}x\right)\cosh\left(\sqrt{3e_0}a\right)}e^{\mp\sqrt{3e_0}\coth\left(\sqrt{3e_0}a\right)x}$$

$$= \left(1 \pm \tanh\left(\sqrt{3e_0}a\right)\tanh\left(\sqrt{3e_0}x\right)\right)e^{\mp\sqrt{3e_0}\coth\left(\sqrt{3e_0}a\right)x}.$$

$$\text{(A.95)}$$

The corresponding eigenvalues are

$$\lambda(a) = -\wp(a) = -e_0 - \frac{3e_0}{\sinh^2\left(\sqrt{3e_0}a\right)}. \quad \text{(A.96)}$$

Let us study the states of the infinite conduction band. These correspond to parameters a that lie on the imaginary axis in the segment defined by the origin and the imaginary half-period ω_2, i.e. $a = ib$, where $0 \le b \le \frac{\pi}{2\sqrt{3e_0}}$. These states assume the form

$$\psi_\pm (x; a) = \left(1 \pm i\tan\left(\sqrt{3e_0}b\right)\tanh\left(\sqrt{3e_0}x\right)\right) e^{\pm i\sqrt{3e_0}\cot\left(\sqrt{3e_0}b\right)x}. \qquad (A.97)$$

The corresponding eigenvalues are

$$\lambda\,(b) = -\wp\,(a) = -e_0 + \frac{3e_0}{\sin^2\left(\sqrt{3e_0}b\right)}. \qquad (A.98)$$

The range of allowed values of the parameter b implies that the sin in the above formula takes all values from 0 to 1, and thus the energy takes all values from $+\infty$ to $2e_0$, as expected.

These states asymptotically at $x \to \pm\infty$ look like

$$\psi_+ (x; a) \xrightarrow{\;x\to\pm\infty\;} \left(1 \pm i\tan\left(\sqrt{3e_0}b\right)\right) e^{i\sqrt{3e_0}\cot\left(\sqrt{3e_0}b\right)x},$$

$$\psi_- (x; a) \xrightarrow{\;x\to\pm\infty\;} \left(1 \mp i\tan\left(\sqrt{3e_0}b\right)\right) e^{-i\sqrt{3e_0}\cot\left(\sqrt{3e_0}b\right)x},$$

i.e. they are free waves. Actually, this is expected, since the potential asymptotically tends to a finite value. This also implies that these states are delta-function normalizable states. Indeed, the asymptotic wavenumber k assumes the value $k = \sqrt{3e_0}\cot\left(\sqrt{3e_0}b\right)$, which fits exactly what we would expect for a free particle, namely

$$E - V\,(\infty) = -e_0 + \frac{3e_0}{\sin^2\left(\sqrt{3e_0}b\right)} - 2e_0 = 3e_0\cot^2\left(\sqrt{3e_0}b\right) = k^2.$$

The above are not surprising. The surprising fact is that the states, which are asymptotically right-going free waves at $-\infty$, namely the states y_+, are also right-going free waves at $+\infty$. The same holds for the left-going states y_-. Therefore, *the reflection coefficient vanishes*.

Let us now study the appropriate limit of the finite valence band. These states correspond to parameters a of the form $a = \omega_1 + ib$. However, at the limit we study ω_1 diverges. Therefore, in order to study the valence band, we need to find the limit of $\psi_\pm (x; a + ib)$, given by (A.95), as $a \to +\infty$ and $0 \le b \le \frac{\pi}{2\sqrt{3e_0}}$. This is easy to find

$$\lim_{a\to+\infty} \psi_\pm (x; a + ib) = \left(1 \pm \tanh\left(\sqrt{3e_0}x\right)\right) e^{\mp\sqrt{3e_0}x} = \frac{1}{\cosh\left(\sqrt{3e_0}x\right)}. \qquad (A.99)$$

Notice two things:

1. Both eigenfunctions have *the same limit*. We have seen in Chap. 5 that this occurs for states that lie at the edges of the allowed bands. This is necessarily the case in the double positive root limit that we study, since the valence band degenerates to a single energy value. Actually, this is the reason the whole band has the same limit.

2. This eigenfunction is *normalizable*, since

$$\int_{-\infty}^{+\infty} \frac{1}{\cosh^2\left(\sqrt{3e_0}x\right)} = \frac{2}{\sqrt{3e_0}}.$$

Let us now follow the approach of Sect. 5.2.2, in order to investigate the above properties of this special potential. We consider the superpotential $W = \sqrt{3e_0}\tanh\sqrt{3e_0}x$. It is a matter of trivial algebra to show that the two partner potentials are

$$V_1(x) = W^2(x) - W'(x) = 3e_0 - \frac{6e_0}{\cosh^2\left(\sqrt{3e_0}x\right)},$$

$$V_2(x) = W^2(x) + W'(x) = 3e_0.$$

Therefore, if we neglect an overall shift by e_0, which is physically unimportant, the special positive double root limit of the bounded $n = 1$ Lamé potential (A.94) is partner of a flat potential.

We know that if y is an eigenfunction of the potential V_1, then $Ay = y' + Wy$ will be an eigenfunction of the potential V_2 with the same energy. Similarly, if y is an eigenfunction of the potential V_2, then $A^\dagger y = -y' + Wy$ will be an eigenfunction of the potential V_1 with the same energy. This explains the behaviour of the conduction band states and the existence of the single bound state that we found above. If we act with the creation operator on the free wave eigenstates of the flat potential, we will get

$$A^\dagger e^{ikx} = \left(-ik + \sqrt{3e_0}\tanh\sqrt{3e_0}x\right)e^{ikx}, \qquad (A.100)$$

which are the states of the infinite conduction band (A.97) upon the identification $k = \sqrt{3e_0}\cot\left(\sqrt{3e_0}b\right)$. At the limit $x \to \pm\infty$ the action of the creation operator on a free wave is equivalent to multiplication with $-ik \pm \sqrt{3e_0}$ and thus, it cannot transform a right-going wave to a left-going one or vice versa. For this reason the non-trivial potential (A.94) is reflectionless.

But, what about the discrete spectrum of the two potentials? It is a general fact in supersymmetric quantum mechanics that the potential V_2 lacks the energy level of the ground state of the potential V_1. This is due to the fact that the latter obeys $Ay_0 = 0$. Indeed, the normalizable state (A.99) obeys

$$Ay_0 = \left(\frac{d}{dx} + \sqrt{3e_0}\tanh\sqrt{3e_0}x\right)\frac{1}{\cosh\sqrt{3e_0}x} = 0. \qquad (A.101)$$

This is the underlying reason the potential (A.94) has one bound state, unlike its partner, the flat potential.

Printed in the United States
By Bookmasters